The business of ecology

Australian organisations tackling environmental issues

Edited by Leigh Cato

ALLEN & UNWIN

Publisher's note
We have made every effort to produce this book using environmentally-endorsed materials. The text and colour section are printed on chlorine-free paper, and the cover is printed on chlorine-free cardboard. The colour section and cover are printed with soy-based inks, and the cover is coated with a cellulose fibre lamination.

First published 1995
Allen & Unwin Pty Ltd
9 Atchison Street, St Leonards, NSW 2065 Australia

National Library of Australia
Cataloguing-in-Publication entry:

The business of ecology: Australian organisations tackling environmental issues.

ISBN 1 86373 733 2.

1. Social responsibility of business—Australia.
2. Economic development—Environmental aspects—Australia. 3. Sustainable development—Australia.
4. Industries—Environmental aspects—Australia.
I. Cato, Leigh.

658.408

Set in 10/12.5 pt Garamond by DOCUPRO, Sydney
Printed by McPherson's Printing Group, Australia

10 9 8 7 6 5 4 3 2 1

Contents

Abbreviations

ACF	Australian Conservation Foundation
ACMRR	Australian Centre for Minesite Rehabilitation Research
AMIC	Australian Mining Industry Council
AMIRA	Australian Mineral Industries Research Association
CONCOM	Council of Nature Conservation Ministers
CSIRO	Commonwealth Scientific and Industrial Research Organisation
CSU	Conservation through Sustainable Use
DASCEM	Department of Administrative Services Centre for Environmental Management
EIS	Environmental Impact Statement
EMP	Environmental Management Plans
EPA	Environment Protection Authority
ERMP	Environment Review and Management Program
ESD	Ecologically Sustainable Development
GDP	Gross Domestic Product
GNP	Gross National Product
IGAE	Intergovernmental Agreement on the Environment
LEAP	Landcare Environmental Action Program
NAFI	National Association of Forest Industries
NFPS	National Forest Policy Statement
NPP	Net Primary Production
RAC	Resource Assessment Commission
RGG	Renison Goldfields Consolidated
SECV	State Electricity Commission of Victoria

TWS	The Wilderness Society
UNEP	United Nations Environment Program
VOC	Volatile Organic Compounds
WMC	Western Mining Corporation
WMI	Wildlife Management International Pty Ltd
WWF	World Wide Fund for Nature

Glossary

Acid rain Rain made acid by contamination, especially by industrial waste gases.

Aldehydes Compounds produced by oxidation (combustion) of alcohols such as ethyl alcohol, and methyl alcohol (ethanol and methanol).

Algae (sing. alga) Primitive non-flowering photosynthetic plants, mainly aquatic; e.g. seaweeds and many plankton.

Anaerobic ponds Used for sewage purification; unlike aerobic ponds, do not require oxygen.

Bacterium (pl. bacteria) Microscopic single cell plant noted for spreading disease.

Biodegradable Capable of being decomposed by living matter.

Biodiversity All living individual plants, animals, fungi and micro-organisms sustained by the Earth.

Biofuels Fuels that originate from biological systems, such as wood from trees.

Biota The animal and plant life of a region.

Bioremediation Restoration or rehabilitation of the plant life of a region.

By-product Secondary product that results from the manufacture of something else.

Capital Assets (products) that can generate income.

Catalyst Substance, person or thing that causes change.

Carbon dioxide Gas formed by burning carbon or by respiration; considered a very potent 'greenhouse gas', although non-toxic.

Carbon monoxide Gas produced when carbon-based fuels are only partially burned; it is colourless, almost odourless but toxic and generates 'smog'.

Catalytic converter A **catalyst** is a substance that aids or accelerates a chemical reaction without itself undergoing change. A catalytic converter uses the oxygen in the exhaust systems of modern vehicles to convert carbon monoxide to carbon dioxide, and to convert partially burned hydrocarbons to water vapour and carbon dioxide.

Chlorofluorocarbons (CFCs) Synthetic compounds of chloride, fluoride and carbon; used in refrigerators and some aerosols, among other applications.

Closed loop control Control of the fuel and air mixture in a vehicle equipped with a catalytic converter, so that the engine meter supplies the minimum of oxygen required for complete fuel combustion.

CNG Compressed natural gas.

Corporatisation A legal framework by which state-owned commercial operations are obligated to pursue market-based objectives.

Crocodilians Any large tropical amphibious reptile of the order Crocodilia, with thick scaly skin, long tail and long jaws.

Cullet Broken/refuse glass used in glassmaking, usually mixed with new material.

Demand The desire for a particular product or service supported by the necessary means of exchange to gain ownership.

Ecology The branch of biology dealing with the relations of organisms to one another and their physical surroundings.

Ecologically sustainable development (ESD) A broad term introduced in 1987 by the Brundtland Report, *Our Common Future* which defined it as development that meets the needs of the present without compromising the future generations' ability to meet its own needs.

Ecosystem Interdependent system of living plants and animals that can coexist with other different ecosystems.

Effluent Sewage or industrial waste discharged outside the normal disposal system.

EIS Environmental Impact Statement.

Environment Physical surroundings and conditions.

Environmental audit A full assessment of an organisation's interactions with the environment; e.g. equipment, transport, resource use, emissions.

Evapotranspiration Process by which liquids are drawn up through the ground and then evaporated.

Fossil fuel Coal, oil etc. formed in the geological past.

Four Rs Reduce, reuse, recycle, recover.

Global warming The build-up of greenhouse gases; see **Greenhouse effect**.

Greenhouse effect Global temperatures raised by a blanket of gases around the earth reflecting infra-red radiation back to the surface.

Greenhouse gases Gases that reflect infra-red radiation from the Earth's surface; e.g. water vapour, carbon dioxide, methane, nitrous oxides and chlorofluorocarbons.

Greening Restoring vegetation to sustainable land management levels; see also **Sustainable land use**.

Hardwood Wood from trees that are classified botanically as Angiosperms and are mostly broad-leaved; the wood is porous. The term does not necessarily denote the hardness of the wood.

Heavy metal pollution High-density metals not easily disposed of and contributing much to industrial pollution.

HCFC Hydrochlorofluorocarbon.

Hydrocarbons Petrol consists of a mixture of carbon molecules that also contain hydrogen atoms; it is usually is 84 per cent carbon and 16 per cent hydrogen.

Integrated waste management Waste disposal that recycles and recovers products and waste.

Intractable waste Waste not easily disposed of.

Landcare Generic term for the hundreds of rural groups that have formed over the last decade; also used to describe government-funded programs helping them.

Leachate Any toxic substance or compound that leaches into the ground or watertable, often as a result of incorrect disposal.

Lead in petrol Lead added to the base petrol formula in the form of tetraethyl lead increases its octane rating. Lead is toxic and implicated in intellectual development impairment, particularly in children.

LPG Liquid petroleum gas.

Mega-diversity A great extent of biodiversity.

Municipal waste Includes putrescible and solid waste from manufacturing, commercial, processing and service industries and households but not liquid waste, sewage or greasetrap waste.

Nitrate A component of fertiliser; any salt or ester of nitric acid.

Nitrification The impregnation of a substance with nitrogen.

Non-renewable resource A depleted product source such as oil, coal.

NPP Net primary production.

Octane levels Octane, a component of petroleum, aids energy efficiency and is essential for engine maintenance.

Organic waste Any waste-stream substance with the capacity for life or growth; e.g. waste fertiliser, chemical by-products.

Overburden Rock that must be removed before the mineral deposit beneath it is mined.

Oxides of nitrogen (NOx) Products of any fuel using air as the oxidant for combustion; nitrogen represents about 79% of air by volume.

Ozone depletion Depletion of Earth's protective atmospheric layer as a result of environmental degradation, global warming and greenhouse effect.

PET Petalite; plastic soft-drink bottles made from polyethylene terephthalate.

Pollutant Any substance that contaminates the environment.

Process effluent Effluent resulting from manufacturing or treatment processes.

Process emissions Emissions given off in production or treatment processes.

Property rights Rights of ownership. For economic efficiency they need to be private, defined, monitored and tabled.

Pulp log Logs unsuitable for sawmilling, used to produce woodchips, pulp, paper, and wood panels.

Rehabilitation Restoration to normal or previous status.

Residual pollution Pollution that remains even after the source has been eradicated.

Salination High levels of mineral salts in the soil caused by rising watertables. There are two types: (1) dry-land salination that occurs when trees and vegetation are cleared, thereby increasing water run-off and raising watertables downstream; (2) wet-land salination caused by irrigation.

Saw-logs Logs suitable for sawmilling.

Slag residue Substances left in or on the ground by industrial plant slag-heaps.

Sludge Greasy, muddy or slushy sediment.

Softwood Timber from conifer species such as radiata pine and cypress pine.

Stratospheric ozone versus 'bad ozone' Stratospheric ozone benefits the upper atmosphere by inhibiting ultraviolet radiation but pollutes the earth's atmosphere when it combines with hydrocarbons and nitrogen oxides to form smog.

Superphosphate Fertiliser made by treating phosphate rock with sulphuric or phosphoric acid. Overuse has led to land and waterways degradation.

Sustainable land use The managed use of land to maintain the environment and minimise or reduce damage to it.

Toxin A poison of animal or plant origin.

Tree farms Large, managed estates for supplying trees.

Volatile Organic Compound (VOC) Substances containing carbon that may be formed by the combustion of fossil fuels and evaporate readily when exposed to the atmosphere; e.g. unburned hydrocarbons from vehicle exhaust that act with NOx to form smog.

Waste-stream A flow of waste products; the total collection of domestic and industrial waste.

Wastewater Term for liquid industrial and domestic waste; e.g. liquid effluent, sewage.

Acknowledgements

Very special thanks to Ken Cato for his support and encouragement and to Bridgette Newbury and Barbara Castles for their attention to detail, many hours of word processing, editing and sensible suggestions. I also wish to thank Elizabeth Weiss, my publisher, for her continued support and enthusiasm.

Introduction

LEIGH CATO

> We do not say that a better future happens automatically or without effort. It will happen because men and women—sometimes as individuals, sometimes as enterprises working for profit, sometimes as voluntary non-profit-making groups and sometimes as government agencies—will address problems with muscle and mind, and will probably overcome, as has been usual throughout history.
>
> Herman Khan & Julian L. Simon—*The Resourceful Earth*, 1984

The state of the environment is littered with buzzwords: industrial pollution; global warming; creating a sustainable environment; protecting biodiversity; making businesses ecologically sound; conservation; preservation; ozone depletion; corporate responsibility; green legislation; environmental controls; the greenhouse effect; environmental vandals; environmentally friendly; sustainable development—what's it all about? How bad are the problems? Who is causing them? More important, who is fixing them?

The protection of our biodiversity and the assurance of a sustainable future are becoming fundamental responsibilities for every company and organisation; they are all required to minimise the environmental impact of their operations and their products. In the future, those who ignore their environmental obligations may become uncompetitive; their directors could be liable to litigation over any environmental damage caused by their activities and products. Corporate environmental responsibility is now playing a greater and more integrated role in selling products, disposing of waste, attracting investment, and recruiting and retaining staff.

It is only in the last fifteen years or so—that is, since the effects of global warming, the destruction of the rainforests, and holes in the ozone layer have been widely publicised—that the state of the environment has become a serious, international concern.

In the late 1960s and early 1970s conservation groups in the USA and Europe were largely responsible for alerting the general public to threats to the environment, and lobbying for the first preventative legislation. The conservationists' urgent and highly emotive style was picked up readily by the media and enjoyed widespread coverage. The 'scare tactics' and emotive reporting also had a detrimental effect on the corporate world, which was portrayed as a group of 'bad guys'. Many corporations did not have the communications skills to combat these claims. Alarmed by media sensationalism and unbalanced reporting, many withdrew into silence. In this void, rumours and distortions found fertile ground.

Many companies were responsible for the problems that some of their processes, products, and effluents were causing but were unaware of the consequences. As pressure and awareness grew, companies in the industrialised nations started to respond—some more quickly than others. The more far-sighted organisations began to form corporate environmental policies, governments established tighter legislation, and consumers started to include environmental factors in their decision-making.

Yet, years later, the questions remain: is the industrial and corporate world made up of environmental vandals or are most organisations responsible, doing the best they can with the available means? What have they accomplished? And is it enough?

In Australia the state of the environment compared very favourably with that of many industrialised countries. This is partly due to its relatively late European settlement and a population that is small compared to the land mass. However, according to some statistics, we have experienced significant species loss, salinity problems, pollution of land, air and waterways, and we produce a remarkable amount of garbage per person.

An historical perspective helps us to realise how our current views and perceptions have been shaped. Over the past two hundred years Australia's affluence has been closely tied to the development of agriculture, mining and industry. The land was unyielding, and for early settlers even subsistence farming was a struggle. As the population grew, it made ever-increasing demands on natural resources, and gave little consideration to the environment.

The number of Australian species lost because of human intervention over the years, the amount of land that has been destroyed and the extent of the effects of industry and mining on the environment are in dispute because they are difficult to establish precisely. However, most industrial companies, conservation groups, environmental researchers, governments and industry associations in Australia agree that the state of the environment is enough of a problem to warrant considerable attention and to support the need for changes. These groups tend to diverge when it comes to determining how urgent the environmental problems are, what means of action should be employed, and how far these measures should be taken. Those not directly involved in these issues find it hard to decide which of these competing claims are most important, and hard to understand the dilemmas and obstacles that have to be overcome to solve the problems. For example, a chemical company may have equipment that should be more environmentally efficient, but cannot afford to upgrade it unless another plant is closed down, putting 800 people out of work.

The initial concept for *The Business of Ecology* arose from the realisation that many Australians who are not directly involved in environmental debates know very little about the background and issues at stake. Although we cannot all be experts, many of us would like to be able to understand the claims and issues within their context.

How many Australians have the impression that the mining industry is committed to the rehabilitation of mine sites? Which manufacturers are cleaning up their industrial effluents? Does recycled paper have as much conservation value as virgin plantation paper? How well-founded are the arguments for allowing people to farm crocodiles in their natural habitat, or to export cockatoos in order to preserve the species?

This book tackles these issues and presents case histories to demonstrate some of the problems and opportunities. It describes some of the steps that are being taken to ensure that the environment we depend upon for our own survival is not destroyed. The criteria for determining which companies and organisations would be approached for inclusion in the book were based on their significant, tangible environmental achievements. Accredited conservation groups, industry associations, and those who bestow recognised environmental awards were asked to recommend good examples of corporate environmental responsibility. The information in the case histories has been supplied by the companies and represents their views. These organisations give a cross-section view of what is going on in a great many of Australia's companies and government bodies. They are not all beyond reproach, environmentally speaking, but they have all made genuine efforts to integrate environmental policies into every aspect of their businesses. In many cases this has led to developing these policies beyond senior management level to encompass environmental audits, long-term environmental management principles and objectives, and the creation of an infrastructure and monitoring system to effectively ensure their success.

For many of these companies and government authorities it has been difficult to find a forum in which to communicate their efforts and technologies. Many of the case histories include some historical background or industry perspective as a context for the problems and challenges that these organisations have had to face. The case histories cover everything from cradle-to-grave recycling, product design, and future methods of waste management. The different approaches demonstrate what can be achieved in all phases of production: from raw materials to the end product and any residual materials that require recycling, treatment, or disposal.

Environmental issues seem to provoke a plethora of strong views and opinions. Essays by various conservation groups and industry representatives have enhanced the book by contributing a range of approaches taken by Australian private and public organisations in their efforts to achieve environmental responsibility. The information and opinions put forward in the essays and case histories were presented by the organisations themselves.

SECTION 1

Questioning ecological assumptions

The original inspiration for The Business of Ecology *came from a dinner for prominent business leaders in Melbourne at which Dr Grahame Webb was the guest speaker. The topic of his speech was 'The wise use of wildlife'. I had anticipated doom and gloom, the ills of alligator handbags and disappearing rainforests; I was prepared to nod my head diligently, being not only the proud owner of a compost bin but a faithful recycler, and a regular contributor to several conservation groups. However, my repertoire of 'facts' about ecological issues did not prepare me for his passionate, interesting speech, and I feel it is appropriate to start the book with it, to challenge our common assumptions about ecology.*

1.1

Confronting preconceptions and developing insights

DR GRAHAME WEBB, DIRECTOR OF WILDLIFE MANAGEMENT INTERNATIONAL PTY LTD

Dr Grahame Webb is the Director of Wildlife Management International Pty Ltd (WMI), a wildlife research and management consultancy firm established in 1978. For many years the firm was devoted largely to crocodile research and management, and one of its most important clients was the Northern Territory Government, but over the last five years the scope of WMI's activities has increased greatly. Based in Darwin, WMI now provides consultancy services throughout Australia and Asia, including Japan, the Philippines, Indonesia, Thailand, China, Singapore and PNG. Its main concern is still wildlife management, particularly the extension of technology into Asian countries where wildlife is still treated as an economically valuable, renewable resource.

WMI is also involved in the conceptual and philosophical side of complex issues such as endangered species management, trade in wildlife, the sustainable use of wildlife, and sustainable development in general. Their latest initiative is the construction of an International Crocodile Research Centre in Darwin which is intended to become a significant centre for conservation, research and education, and for the development of production technology for the world's crocodilians.

Grahame Webb begins by saying that we might at first mistake him for an environmental vandal instead of a passionate environmentalist. But as he explains, he has found good reason over the years to question the philosophies upheld by the environmental movement—particularly those to do with wildlife conservation. In some areas, he believes, we are headed down the wrong path. He also recognises that a great many environmental issues are based on a relatively few common assumptions.

In my view, the relationship between humans and the natural world varies greatly from culture to culture. There is not necessarily a right relationship, nor a wrong one. There is no one set of ideas or philosophies that can, nor should, be expected to suit all people. Neither should one be imposed on all. Humans have evolved as part of the natural world with an evolutionary obligation to use plants and animals. Most societies and cultures adopt a role, to varying degrees, as guardians or protectors. However, it should be remembered that there is no universal obligation to conserve things for their intrinsic value per se. By definition, if any benefits are derived from the efforts to conserve, the motivation becomes instrumental.

I believe biodiversity conservation would be far more successful if it were based on the potential human 'uses' of living organisms. There is not a living plant or animal whose conservation could not be justified on the basis of its instrumental value. Humans, like all living organisms, affect their environment and consume other forms of life. If humans and other living organisms do not use other living materials they die. It is a matter of basic survival, and the right of humans and of all species to use other living organisms is unequivocal.

Only about 7 per cent of the total Australian workforce is involved in primary production. Most Australians live in a relatively few large coastal cities. Here, amidst high-density living conditions, daily survival no longer depends on the family having to harvest the land, the rivers or the sea. Instead, survival depends largely on purchasing food from the shelves of the supermarket, filled to capacity with produce that has often been reaped from the environments of distant lands and peoples.

It is not surprising that city people (i.e. most Australians) 'value' wildlife in a somewhat intangible way. It is out of proportion to both calorific value as food, and its uses for other practical purposes. Since we were children, city people have accepted the dogma that all native animals are 'good'. There are few avenues open through which they can reach this conclusion from personal experience. The days when the value of wildlife was imprinted on children as they snuggled into the long fur of a wild animal skin, are gone—the fur replaced by acrylic carpet, made from non-renewable resources.

I believe that questioning and critically educating environmentalism is an important step towards developing informed, objective attitudes to environmental issues. It does not mean denigrating it, nor detracting from the vital role that environmentalists have played in making us environmentally aware. The following section identifies thirteen issues which I believe cause a good deal of confusion in the community.

1 CONFRONTING FUNDAMENTALISM

The first issue is 'environmental fundamentalists'. These are people who have adopted an almost religious view of their chosen field, with a set philosophy they are not prepared to change. Unfortunately, for these people the philosophy is in danger of becoming all-important, just as it has done with many religions. Tens of millions of people have died in religious wars over who has the right philosophy of living in harmony together. Could people ever get to the stage of killing people over environmental issues? It is frightening to consider that some animal liberationists have actually blown up laboratories.

2 UNDERSTANDING OUR TWO BRAINS

The second issue concerns our 'two brains'. Humans have evolved steadily over the last million years, yet it is only in the last 10 000 years that agriculture and the domestication of animals has occurred—a mere blink of the eye in human history. As Paul Erhlich suggested, we essentially have an 'old brain' still responding to the basic survival issues associated with 99 per cent of our evolution, and a 'new brain' that tries to respond to the current situation. Our fundamental interest in wildlife, and indeed in conservation generally, may be an 'old brain' phenomenon. The media attracts our attention by appealing to our 'old brain', although the issues themselves are often 'new brain'. There is a continual conflict going on which we do not fully understand.

3 UNDERSTANDING THE ROLE OF THE MEDIA IN ENVIRONMENTAL ISSUES

The media is always being blamed for one thing or another, too much or too little. It is certainly the main battleground for environmental debates, and is the means by which people end up with strong opinions about issues in distant places even though they seldom have any first-hand experience of them.

The power of television, from a biological point of view, is simply astounding. For the first time since humanity evolved, people replace activities directly related to survival with watching a small box emitting images and sounds. Add to this the sound biological reasons why we should be more interested in bad news (which can kill you) than good news (which you can wait until tomorrow to hear) and you have a problem. The 'old brain', which long ago learned to believe what one 'sees', searches out (by changing channels) the 'bad' news that is stimulating. Thus although we frequently blame the media, it is really our problem—we demand the 'bad' news and are becoming less and less satisfied with the distorted picture of the world that we subsequently perceive.

The significance to environmentalism? Take the mining industry as an example. An environmentalist decides a particular mine may adversely affect a particular animal species—it may drive it to extinction in that area. Such news fires the 'old brain' biologically—we do not want to see animals become extinct because such losses were once intimately related to survival. Bingo! It's on national television, beamed to households across the nation. In contrast, the rehabilitation of 100 hectares of mining land in the Great Simpson Desert does not fire our hunter-gatherer spirit. In short, the mining companies want to project 'good' news, but we are not prepared to listen to it. Advertising, at enormous expense, is the only option open to mining companies attempting to dispel environmental myths. But even this is clearly an advertisement—it is not the 'impartial' judgment of the news broadcaster.

4 QUESTIONING DOOMSDAY ATTITUDES

The fourth issue is the 'doomsday' approach to the environment. I truly believe that the environmental 'bad' news is having a dramatic effect on all Australians—especially the 80 per cent that live in cities well separated from the bush. If you ask young children who live in cities to write stories about the future you are likely to get all sorts of bizarre answers: no trees, no animals, deserts everywhere, thick smog, foul-smelling air, and so on—the result of all the 'bad' environmental news. Strangely, we then seem shocked by abuses such as the use of heroin—'Why are you burning your bridges Johnny? Why aren't you planning for the future?' We spend most of our time telling Johnny that there will be no future.

5 ACCEPTING THE FACTS ABOUT KILLING

'Killing' is a highly sensitive issue, partly because we relate the killing of animals to the killing of people. For people to kill people, there must be some hatred. But does this mean you must hate any animal that you kill? Nothing could be further from the truth. Most hunter–gatherer societies, especially Australian Aboriginals, kill animals continually—but their love and respect for animals is embodied in every aspect of their culture and religion. It is the hunters who have a real vested interest in the conservation of their prey.

Killing and conservation are two totally separate issues. People who oppose the killing of animals, for any purpose, adopt a philosophical position—which is clearly their right. But it may not have much to do with conservation and in some cases is in conflict with conservation. The pragmatic attainment of conservation goals may require the killing of hundreds, thousands and even millions of animals. If we could kill all cane toads, rabbits, feral cats and foxes in Australia, many conservation problems would be solved. In fact the conservationist who devises the best method of killing these animals may well go down in history as Australia's 'Conservationist of the Century'.

6 SOLVING ENVIRONMENTAL PROBLEMS

If we have a medical problem, such as an ingrown toenail, we expect the person charged with solving the problem to have had six years of training and a proven track record. Yet there are 5.5 billion people in the world, which makes about 55 billion toes and toenails. In the overall scheme of things, one ingrown toenail is a trivial problem. When it comes to environmental problems, which are often immensely complex and can involve the welfare of thousands of people, to whom do we turn for problem-solving advice? All too often it is to people with a burning passion for the subject but no formal education or training in that particular field.

7 Establishing priorities

There is no way of ordering the relative importance of environmental problems. Thus, although we may consider a vast oil spill to be more important than an annoying possum in a neighbour's tree, both are environmental problems and there is no objective way of saying why one is more important than the other. The consequences are significant. Environmental debates often involve identifying as many potential problems as possible, so that the sheer weight of numbers carries the day. The person charged with solving the problems can never get to the end as more and more potential problems can be generated with ease.

8 Balancing production with biodiversity

The eighth issue is one that concerns the topic of this book—maintaining production versus maintaining biodiversity. The world is a revolving globe which collects solar energy. This energy, collected in green plants, drives all life on Earth. The economies of most countries are related to the energy-collecting capacity of their lands and territorial seas. Problems affecting the productivity of the land are more important than those affecting biodiversity. We can replace native forest with crop land and still survive, but we have real problems if the land becomes unproductive regardless of whether we wish to use it for native forest or agriculture.

9 Defining practical strategies for biodiversity

What we really need are practical strategies for preventing the loss of biodiversity. Concentrating on problem awareness leaves the impression that once people are aware, the problems will somehow solve themselves. For example, we are increasingly being made aware of serendipitous findings; a previously insignificant animal or plant is suddenly, by chance, found to contain something that is of use to people. It is a good argument for selling biodiversity to the public. However, the implicit assumption is that if you save the animal, you will be able to use it. Most people promoting biodiversity awareness are totally committed to not using wildlife.

10 Habitat protection

How much of the existing diverse habitats can we protect? Currently about 5 per cent of the world's land surface is in protected areas of one form or another. But if all the conservation dollars available were to be spent on that 5 per cent, it is still doubtful whether we could really preserve them. The option of doubling the number of protected areas is simply not available. Land is a country's fundamental asset and the energy trapped in it must be converted to food and commodities if people are to survive and/or maintain their quality of life. Our survival and biodiversity in general are going to depend on what happens outside protected areas—in the 95 per cent of land not nominated for preservation. Clearly, a conservation priority should be the development and testing of new and innovative strategies for protecting habitats on lands outside protected areas—strategies that do not conflict with the need to use this land for economic gain.

11 EVALUATING THE COMMERCIAL SUSTAINABLE USE OF WILDLIFE

The conservation advantages of wildlife use is a very controversial area. But in many cases it offers the most viable solution for raising the value of natural habitats and creating tangible incentives for their protection. In Australia the low price of fox fur decreases hunting rates, increases the fox population, and in turn, increases the negative impact of foxes on native wildlife. Foxes and cats may well be the prime causes of mammalian extinction in Australia. Does it make conservation sense to oppose the use of fox furs? It would seem to me that such opposition is conservation vandalism.

12 WILDLIFE AS A RENEWABLE RESOURCE

Whilst animals may look like exquisite, unique paintings, they are renewable. They can be reproduced over and over again given the right conditions. Similarly, animal populations have various means of compensating for losses in numbers due to mortality or hunting. For example, if 1000 hatchling saltwater crocodiles enter the river, 100 may survive for one year. If only 200 enter the river, 100 may still survive.

13 'AUSTRALIAN HYPOCRISY'

The thirteenth issue encompasses what I refer to as 'Australian hypocrisy'. In the outback of Australia it is possible to purchase large tracts of land and degrade their productive potential through conventional agriculture. Clearly we need agricultural production, and there will always be an environmental cost associated with it. However, there is no reason why vast tracts of land could not be used for wildlife production and eco-tourism, with the possibility of ploughing the profits back into habitat management so that more wildlife can be used for commercial gain. Yet this approach is typically frowned upon. When pressure does mount in Australia for trade in wildlife, the option of captive breeding is all too often seen as the safe way to go. Yet captive breeding establishments are isolated from the wild, and thus the profits made are unrelated to habitat protection. In its simplest form, the species of concern can become extinct in the wild, while income earned from its survival in captivity increases. The maximum conservation advantage is accrued when wild harvest programs create commercial reasons for ensuring that the target population and its habitat remain intact, and rehabilitated or improved.

These issues are some of the many that I believe need to be questioned. Environmental awareness and responsibility must become part of our daily lives, but we must find ways of doing this without massive cost to the community. It makes no sense to squander resources on any problem.

1.2

Marketing the environment by adding value

PATRICIA CASWELL, EXECUTIVE DIRECTOR OF THE AUSTRALIAN CONSERVATION FOUNDATION

The Australian Conservation Foundation (ACF) describes itself as having over 20 000 members, with another 90 000 supporters and donors, recipients of its catalogue and a staff of over 60. Its national headquarters are in Melbourne, and it has offices in New South Wales, Western Australia and the Australian Capital Territory. The ACF works on campaigns, research, policy development, marketing, membership services and public awareness. Volunteers also provide a vital resource—their talent and time.

Over 80 per cent of the ACF's funding comes from membership fees, donations, publications and the sale of ACF products. The rest comes from government and other grants which are usually project-based.

According to Patricia Caswell the ACF has been working for ecological sustainability for 25 years. As a watchdog of government and industry, ACF claims to have been instrumental in securing public participation in the political process and has succeeded in putting the environment high on Australia's agenda.

Public awareness and environmental education are ongoing priorities, as are the various campaigns which come under three broad action areas: global change; biodiversity; and sustainable use of natural resources. The most recent undertakings of the ACF include the development of stringent guidelines for corporate memberships, product endorsements, corporate sponsorships and recognition of 'Green Leaps Forward' by industry.

Globally and in Australia we still suffer from the residue of that deeply mistaken view that nature is the enemy of civilisation, a force to be conquered, or a luxury for those urbane enough to appreciate it. The environment is still not counted in the international accounts. Trees, water and wildlife have been free goods. They are only counted in a nation's books when they are so damaged that they need repair—and it costs.

Polluted air is in the accounts because it costs millions to purify it. Damaged river systems, like the Murray–Darling, are on the books because they cost billions of dollars to restore. The United Nations is piloting new methods for counting ecology in the international accounts.

There is a growing recognition that we can no longer regard 'green' as the protection of trees and penguins on one hand, while the development and creation of wealth and jobs continue on the other hand. Australia has been slow to pull together the environment and the economy. It is still immensely difficult to convince our politicians that this is the only way to plan a decent future.

In a major national poll, Australians declared that the environment remained a more significant issue than the Australian balance of trade figures, the health system, taxation, industry, resource development and education. The average Australian is aware of environmental problems. He or she may not necessarily have the right information or the best solution to the problem, but there is certainly the 'will to change'. For example, a study by ANOP Research Services showed that over 70 per cent of those interviewed believed that recycling glass and paper was very important to the environment.

What are the dynamics of conservation in the future? Can we expect governments to take environmental policy seriously when faced with a poor economic climate? In the past, green politics have tended to fade under pressure. In recent times we have heard many criticisms of Australia's scientific commitment. In Australia we spend very little on research. We spend even less on ensuring that brilliant research is translated into technologies for saving the environment. This area has huge potential, but it seems we have understood even less about the science of the environment, ecology and the translation of their needs into technology, than other industrialised countries.

One of the major aims of the ACF is to ensure that Australia will be an increasingly export-oriented nation with a reputation for higher environmental standards, clean production and energy efficiency. To accomplish this we need industry to adopt and develop more effective environmental products, production and waste disposal methods and technologies. This is also part of what are sometimes called the 'brown issues' such as urban planning, design and living which are interrelated with all environmental issues, and from our perspective they are just as green as forests.

The four Rs remain critical habits and we can all adopt them in our households, in our workplaces, in our influence on public policy, and in our communities. We should be reducing our consumption. We should be reusing whatever we can.

We should be recycling and recovering as many products as possible. We believe this can happen in all sectors of the community, government departments, and mainstream industry production lines. To promote this we have developed several marketing strategies that are based on adding value to companies and organisations who are striving to become greener.

THE REDHEAD GOES GREEN—A 'GREEN LEAP FORWARD'

In some cases it means acknowledging a company's effort, and the ACF does this in several ways. One is what we call the 'green leap forward', and the Australian company, Bryant & May, has made this leap. In an initiative to create a Redheads match head that eliminates environmentally damaging chemicals, the company reviewed the entire life-cycle of its match products to produce a head, stick and box that eliminate harmful substances without affecting quality or performance. They also reduced their sources of pollution and switched to recycled materials.

It could be said that the reduction of the use of pollutants and trees is fairly small. However, about 5 billion matches are used in Australia each year and if every company adopted similar practices, the results overall would be substantial.

THE GREEN AUDIT—CRITERIA FOR ACF ENDORSEMENT

We have created corporate memberships for companies that comply with our Corporate Assessment Criteria: companies that are prepared to be scrupulous and believe that environmental responsibility adds value to their products and services. The stringent set of guidelines in the Green Audit ensures that these companies are what they say they are. Consumers can be sure that an ACF Corporate Member is environmentally responsible. There are also companies that have certain products endorsed by the ACF if those products meet the ACF's high environmental standards. This endorsement is for the product itself, not the entire operation of the company.

GREEN JOBS, GREEN FUTURE

Another particular concern for the ACF is the development of green jobs in industry, especially in the areas of environmental protection, rehabilitation and maintenance. Research among employers (70 per cent from the private sector) has shown that there has been substantial and continuing growth in green jobs. The ACF argues for a strong, interventionist industry policy, government assistance in developing of 'green' trade and export markets, and additional infrastructural investment in areas such as housing and transport. Our Green Jobs Unit, set up with resources from the Federal Labor Government, is continuing its expansion throughout a number of industries.

1.3

Rational, not sensational, economic strategies for ecological issues

DR ALAN MORAN, RESEARCH DIRECTOR,
TASMAN INSTITUTE

The Tasman Institute was founded in 1990 and provides consulting and research services to government, international agencies and businesses in a wide range of disciplines including economics, ecology and the law. The overall aim of the Tasman Institute is to explore ways of improving the outcomes of major community issues by making greater use of private enterprise and by reducing regulation.

Dr Alan Moran supports long-term strategies. In the following article he suggests that pragmatic long-term strategies are necessary to reconcile environmental and economic issues.

ECONOMICS, CHOICE AND CONSERVATION

Industry, government authorities and environmentalists generally agree on the need for measures to prevent the extinction of species, especially where extinction results from human actions. However, their opinions often differ about just how much needs to be done and at what cost.

Like so much that confronts our decision-making, the different options must have an economic dimension. Economics places a worth on things according to how people value different goods and services (including environmental services). That worth rests on the trade-offs people make—for example, trade-offs between more work and increased leisure. Such choices reflect people's different needs and desires. Because everyone has different tastes and incomes all of us have unique valuations of goods and services. Adding all these valuations together and setting them against costs of supplying the goods and services enables a price to emerge.

This price is what market-based economic analysis revolves around. While markets do not capture all of the values of goods and services, the best trade-offs can usually be brought about without government interference in the marketplace for goods and services. Moreover, in some circumstances a diminished government role can benefit conservation. Many government actions distort demand and supply by, for example, forbidding certain activities. The former Commonwealth Environment Minister, Barry Cohen, has discussed the various bureaucratic impediments that forced him to abandon his plans to create a wildlife sanctuary–ecotourism facility. Removing restraints on use enables natural resources to be valued and used to their best advantage.

As well as involving trade-offs, income growth and the conservation of flora and fauna can also be mutually supportive. For example, conservation can mean maintaining tree cover to prevent soil erosion, or preserving indigenous predators so that a balance is kept between them and pests. It can also mean restoring and maintaining a more pristine environment to take advantage of opportunities for ecotourism and other leisure needs.

Higher incomes, however, also mean more resources being used, and more people moving about in areas that had previously been considered 'untouched'. In this way, increased pressure on resources is likely. Even so, where there is conflict between income-earning land and conservation, wealthier economies such as Australia are more likely to take on conservation issues. This is because richer societies can more readily afford to improve their lifestyles.

THE COSTS OF CONSERVATION

There are several readily identifiable costs incurred by conserving biodiversity. These include losses due to the denial of a preferred land use; additional expenditures made necessary in the course of, or following, an activity involving land modification; requirements to engage in research as a condition of receiving permission to proceed with development; and government spending on its regulatory agencies and research.

There are three other areas of costs which are not so easily identified but which may be of greater long-term importance: (1) delays in commencing an activity, caused by the necessity to obtain approvals from different levels of government; (2) opportunities lost because of those delays or because, in anticipation of further delays, an application is not advanced; (3) the loss of future commercial activity through the permanent denial of access to unknown, but potentially valuable, land-based resources.

THE BENEFITS OF BIODIVERSITY

The benefits derived from biota and enriched by biodiversity encompass a range of goods and services that are ownable and marketable, including food, water and clothing. The provision of these is, in the main, best left to private ownership.

Biodiversity also comprises a variety of other values that are less securely provided for in an ordinary market system. These encompass recreational and aesthetic values, the values associated with preserving biodiversity for its own sake, and the ethical values asserting that species themselves have a right to exist independently of their benefit or value.

INCOME, MARKETS AND THE ENVIRONMENT

Increased wealth has had a dual influence. It has made the continued existence of natural environments more affordable, yet has posed new threats to the environment. One of the consequences of these opposing tensions is that injudicious approaches to conservation policy have tended to cannibalise the wealth that makes conservation a practical option.

It is necessary to seek out approaches that neither undermine wealth nor destroy things of genuine value to individuals: pragmatic strategies that provide ways of efficiently allocating goods and services—including environmentally valued services—between different and often conflicting needs. Fundamental to these approaches is a more comprehensive and more securely vested system of private property-rights to permit market systems to be fully effective in environmental matters. Without a secure, fully defined system of property rights, people have a natural tendency to socialise their losses and privatise their gains. Property-rights approaches operate by giving land managers and owners incentives to combine conservation and commercial land use; they also encourage them to profit from finding less expensive means of meeting their goals. By combining demand and supply in these ways, private property bases of land use applications best reflect both community and private valuations of biota.

GOVERNMENT DECISION-TAKING AS AN ALTERNATIVE TO MARKETS

The recent increase in concern over the environment is reflected in government land policies. In the past, governments conferred unowned land on individuals on condition that they 'improved' it by clearing and planting. Consequently, native biotic communities and undisturbed environments diminished. Governments now tend to discourage further development of wilderness areas. In doing so they are responding to a 'political' market but the 'battles' this often entails—like those in the buildup to government decisions on Fraser Island, the Queensland rainforests and the Franklin Dam—are ultimately an unsatisfactory means of resolving conflict over land use alternatives.

Unlike the market system, governments have neither the information about true preferences, nor a rigorous incentive system to operate at the lowest cost. The political framework that public management uses for setting and enforcing general policies has likewise proven unsuccessful in allocating resources. While attempting to replicate markets with cost/benefit techniques offers a means of applying some analytical insights, these techniques can never capture the richness and subtleties of demand and valuations that the commercial market marshals. They are therefore, inadequate as policy bases. For this reason, avenues allowing markets that are based on property rights should be employed as comprehensively as possible, before the political determination of resource use is contemplated. In situations where collective ownership of the environment is unavoidable, cost/benefit analysis must be married with policies that incorporate the incentive systems embodied in markets to ensure efficient arbitration between different goals.

EFFICIENT ENVIRONMENTAL POLICIES: THE BASIC INGREDIENTS

Before any improvement in efficiency can be gained by a conservation policy that employs market instruments and secure property rights, certain basic principles must be acknowledged. These include: (1) a minimal disturbance of people's freedom to use their property as they choose, with compensation when their choice is refused; (2) a ceiling on government conservation funding—one that reflects the community's preferences; (3) using public funds efficiently by employing price mechanisms, by opening up trade in conservation assets (native and exotic flora and fauna), and by encouraging competition for the stewardship of environmental assets.

From these basic principles a specific framework to formulate an efficient, market-based conservation strategy could be developed. For the strategy to work, the rights involved must be broadly defined, and include not only land ownership, but leases of public land for grazing or timber harvesting. Where appropriate, legal mechanisms such as licences, easements and covenants can be used. These would make it possible to exchange one asset for another of higher conservation value. It would then be possible to maintain a system of parks, reserves and covenants which maximises the ecological value of community conservation resources. At the same time, land with low conservation value but high commercial value would be put to its most profitable use.

To allow for the 'public goods' value of biodiversity, different gradations of government intervention are necessary. The interventions may be aimed at maximising an agreed biodiversity index or at a well-defined set of conservation priorities. The priorities would have to meet a restricted budget.

SPECIFYING A MARKET-BASED POLICY

People's willingness to pay for the biodiversity of public goods must be the criteria for earmarking funds for conservation, but since this is difficult to quantify we may have to settle for a lesser degree of political accountability by explicitly identifying conservation expenditure in budget papers. Under such a market-oriented approach, possible government conservation interventions can be loosely categorised as falling between minimal intervention and something closer to total direction and control. One end of this spectrum could encompass measures like subsidies to conservation activities on private lands, or franchising the management of reserves and national parks to private conservation groups. At the other extreme there could be continued public ownership and management of national parks and reserves through a partially 'corporatised' conservation trust, which could receive an annual grant, and operate in the same way as national museums.

The trust would trade in conservation 'assets', with its decisions designed to reflect the community's preferences and their differing costs. In limited circumstances, a government conservation trust would be allowed to make compulsory purchases of conservation assets to overcome the problem of hold-outs by landowners sitting on strategic parcels of land. However, full compensation should be required for owners whose property values are reduced by political override. In this way, the vital building-block of prosperity—secure property rights—is not undermined, and at the same time the full extent of the costs of political action is recognised.

It may also be preferable to localise a certain amount of public conservation expenditure by having it borne by the communities that benefit from it through area-specific rates, taxes or charges. Where the communities themselves reap the major benefits, this may allow a more direct expression of individual preferences than is possible with a national body. Alternatively, conservation expenditures can be financed by park entry, charges, franchising kiosks, etc.

Of course, informational and political deficiencies would always conspire to bring imperfections to market-oriented arrangements. Nonetheless, better means are needed for resolving the highly charged political conflicts that have characterised environmental choices over recent years. A policy program, incorporating market-based strategies that can assess the need for conservation measures and promote the survival of species, could help diffuse the tensions that otherwise will surely be amplified in future years.

1.4

Leave nature alone—an evolutionary view of biodiversity

KARENNE JURD, DIRECTOR OF THE WILDERNESS SOCIETY

The Wilderness Society (TWS) began as the Tasmanian Wilderness Society, which was formed in 1976 in response to the damming of Lake Pedder in Tasmania in the early 1970s and the threat of the same thing happening to the Franklin River a few years later. According to TWS, the Franklin River was saved in 1983 as a direct result of the society's mobilisation of public support. Since the success of the Save the Franklin campaign, millions of hectares of magnificent wilderness have been saved through The Wilderness Society's efforts. It is the only conservation group in Australia that campaigns solely for the protection and preservation of wilderness. Successes range from obtaining World Heritage listings for the Daintree Wet Tropics area and Kakadu, to preventing of mining in the Kimberley.

At the time of writing, The Wilderness Society has about 14 000 members across Australia and some overseas, with 23 branches in Australia, and its main campaign concerns are: forests; outback arid and semi-arid wilderness; the Cape York Peninsula; federal and state wilderness legislation; international wilderness issues and wilderness education. TWS is a grassroots organisation whose activities include direct action, education, political lobbying, networking, public outreach and publications. According to Karenne Jurd: 'There are few such strong and invigorating tonics as the peace and solitude which come from watching the sun set over the horizon of the vast western plains of New South Wales. This may seem esoteric and superfluous to some, but it is a powerful gift of the rare and relatively fragile wilderness. A valuable gift that The Wilderness Society believes is well worth cherishing.' Her essay explains the philosophy and undertakings of TWS.

The strategy The Wilderness Society has adopted to tackle the issues threatening the last few remaining tracts of wilderness is the continual lobbying of governments, business and the community alike to protect these vanishing areas. There is plenty that governments can do; the question is, do they have real political commitment?

The Convention on Biological Diversity, signed by more than 150 countries in Rio de Janeiro in 1992 and ratified by the Australian Government, came into effect on 29 December 1993. The Convention recognises that '. . . the fundamental requirement for the conservation of biological diversity is the in situ conservation of ecosystems and natural habitats and the maintenance and recovery of viable populations of species and their natural surroundings'. The emphasis on natural habitats and the natural surroundings of species is important. The Convention provides clear directions (which this country's state and federal governments should take) to finalise a workable national strategy for maintaining the rich biological diversity in Australia. This includes the development of a comprehensive national system of protected areas.

These protected areas would have a vital role in retaining Australia's status as one of the few major regions of mega-diversity in the world. These areas would have to be more than just ecologically viable; they should fully represent the whole range of ecosystems. Australia is the only developed country to host such diversity. These protected areas would be a valuable asset; a unique storehouse for diverse species and ecosystems, many of which are rare or absent in other parts of the world. To this end, it is crucial that the public recognises the importance of the swift completion and subsequent maintenance of the National Wilderness Inventory, and its continued position as a 'key indicator in the development of the reserve system'.

As long ago as 1986 Commonwealth and state governments agreed through the Council of Nature Conservation Ministers (CONCOM)[1] on the need for an active program of Wilderness Area identification and reservation. However, only South Australia and New South Wales have wilderness protection legislation, and even that is not being used effectively. At the time of writing, some states have still not made any appreciable provision for wilderness identification and protection.

Our wilderness areas, which are relatively remote and essentially natural, are large. Without these expansive boundaries, these areas cannot hope to offer any real remoteness, nor to retain their naturalness. For the protection of species and ecosystems, these tracts of wilderness offer the best opportunity by far, because they provide greater resistance to outside pressures than smaller reserves can. In the future, these same wilderness areas will provide the most secure havens in which species and ecosystems can adapt to changing conditions such as climatic changes, or other human-induced problems.

Worldwide, less than 20 per cent of our remaining wilderness is protected. Only Antarctica, the former Soviet Union and Canada have more remaining wilderness than Australia, yet, of the 159 Australian wilderness areas identified by Prineas et al[2] in 1986, only fifteen of them have legislated protection as wilderness. By comparison with the USA, where 37 million hectares are protected under the USA Wilderness Act of 1964, the fifteen protected wilderness areas in Australia amount to only about half a million hectares (0.07 per cent of the Australian continent).

Despite all the political rhetoric and promises, few states are taking any significant action on wilderness protection, and the Federal Government has not demonstrated leadership in this area. Part of the problem is that too few decision-makers recognise the real value of the wilderness. These remote and natural places are not merely places of respite for a few hardy bushwalkers: they are the last living places where select and diverse species, biological communities and ecosystems can evolve and flourish. This in itself offers a wide range of both tangible and intangible benefits. Wilderness areas protect habitats of threatened and endangered species; ensure and maintain the quality of our scarce fresh water supplies; act as buffers against climatic change; and provide a standard against which we can scientifically assess our impact on the rest of our planet. Wilderness areas have recreational and economic values, but they are also important scientifically, psychologically and culturally, and represent a valuable part of our natural heritage. Beyond all this, wilderness areas have an intrinsic value which cannot easily be measured. This too is irreplaceable.

It is interesting to note that the lack of activity on behalf of state and federal government leaders comes at a time when they have made major commitments to protect both our natural and urban environments through the Intergovernmental Agreement on the Environment. Among other things, they agreed that each level of government . . .

> . . . should use their best endeavours to ensure the survival of species and ecological communities, both terrestrial and aquatic, that make up Australia's biota. The parties recognise that the protection and sound management of natural habitats is of fundamental importance to this aim and that all levels of government should use their best endeavours to conserve areas critical to the protection of Australia's flora and fauna and the maintenance of ecological processes that ensure biological productivity and stability.[3]

In the same schedule of the Intergovernmental Agreement on the Environment, Australian political leaders agreed to 'co-operate in fulfilling Australia's commitments under international nature conservation treaties and recognise the Commonwealth's responsibilities in ensuring that those commitments are met'. The state and federal governments have made other commitments and agreements which recognise the importance of wilderness to the protection of our biodiversity, including the National Forest Policy Statement (NFPS) signed by all governments except Tasmania in December 1993. The statement espouses the importance of a comprehensive, adequate and representative network of dedicated and secure nature conservation areas for forests and reserves to protect wilderness.

Not only have the signatories to the NFPS agreed that this reserve system of old-growth forests and wilderness be in place by the end of 1995, they have also agreed that 'until the assessments [of old-growth forests and wilderness] are completed, forest management agencies will avoid activities that may significantly affect those areas of old-growth forest or wilderness that are likely to have high conservation value'.

The question that TWS poses is: Where are the actions to match these words? Old-growth areas, and even areas nominated and assessed as worthy of wilderness protection, are still logged in several states. Activities that affect wilderness are not being avoided, and yet no-one but the conservation movement is lifting a finger to stop this. Indeed, after the Federal Government renewed the export woodchip licences, their commitment to any conservation aims of the NFPS is not worth the paper they were written on. TWS and its members are committed to continuing to speak out in defence of precious and finite wilderness areas. Someone must take governments to task for breaking the agreements they so solemnly made.

It is not that those representing the Commonwealth disagree that wilderness is a key element in maintaining our status as a region of mega-diversity. The Convention on Biological Diversity, to which Australia is a signatory, recognises that in situ conservation of species and their natural habitats is of primary importance. The Intergovernmental Agreement on the Environment acknowledges the responsibility of the Commonwealth to ensure that Australia meets its obligations under international treaties. The Wilderness Society maintains that the Commonwealth has to do more than agree; it has an obligation to protect the remaining wilderness.

Constitutional powers may not permit the Commonwealth to overrule state inaction, but they certainly do provide opportunities for Commonwealth action. With the passing of the tenth anniversary of the High Court's decision which protected the Franklin River in Tasmania's south-west wilderness, there can be few who doubt that the Federal Government can take more action to protect our remaining wilderness, and the biodiversity that it encompasses, than at present.

A discussion paper, prepared in 1992 for the Australian Heritage Commission by consultants Robertson, Vang and Brown,[4] has provided an excellent review of the issues and options for Commonwealth action and the powers that might be used to achieve these objectives. One of the more effective of those options is the introduction of Commonwealth wilderness protection legislation. Proposals for Federal wilderness legislation are not new. In a consultancy report to the Department of Arts, Heritage and the Environment in 1987, John Sinclair[5] proposed the introduction of Commonwealth wilderness legislation. Apart from the protection such legislation would offer to wilderness, the activities associated with its preparation and introduction would greatly enhance [the community's] understanding of wilderness and its values. The enhancement of such awareness of wilderness has generated public debate on legislative reform. This education comes via the grassroots campaign activities of organisations such as TWS.

One example of the Society's ability to stimulate public awareness came in 1987, the final year of a seven-year campaign for wilderness protection legislation in New South Wales. One of TWS's tasks was to generate public awareness, and in nine months hundreds of audio-visual presentations were made to Lions Clubs, Rotary groups, schools, universities, Colleges of Advanced Education, senior citizens' groups, four-wheel drive clubs, a variety of political parties, and to anyone else who would stand still long enough to hear what TWS had to say. There were also speakers' nights, street theatre, suburban letter-writing stalls, bus-stop postcard distribution, telegramathons to MPs and lots of publicity-banner drops from sites around the whole city. Dr Who (complete with Daleks) even came back from the twenty-first century in the Tardis to visit Parliament House in Macquarie Street, Sydney, bringing a message from future generations concerning the dangers of not protecting our wilderness heritage.

TWS is often accused by governments of always being ready to criticise but never praise the solutions. However, when the NSW State Parliament passed the Wilderness Act in December 1987, Premier Unsworth received the world's largest congratulatory greeting card; one that merited a place in the *Guiness Book of Records*.

There is no better way of focusing on the substance of a debate than having to argue for the essential elements of legislation in public. TWS acts as a public conscience, asking questions such as:

- What should be in the code of management for wilderness?
- How can we incorporate the public participation needed?
- How are the interests of Aboriginal Australians to be included from the outset?
- How do we address local landholder concerns?
- Where and how do we deal with compensation issues?
- How can we ensure proper management of the impacts on wilderness values from outside the wilderness area?

At both state and national levels the task of developing public awareness of the need for wilderness protection legislation has not been easy. It is TWS's goal to ensure, through the combined efforts of environment groups, the scientific community and the general public, that our governments, state and federal, make wilderness protection a priority. Without it, Australia and the planet will be the poorer.

Notes

1 Council of Nature Conservation Ministers, 'Guidelines for reservation and management of wilderness areas in Australia', Report by the Working Group on Management of National Parks, Canberra, June 1986.

2 Prineas, P., Lembit, R. and Fisher, N., 'Australia's wilderness: An inventory', Report to the Australian Conservation Foundation, Melbourne, 1986.

3 National Forest Policy Statement, 'A new focus for Australia's forests', Australian Government Publishing Service, Canberra, Dec. 1992; see also Keating, P.J., 'Statement on the Environment', Australian Government Publishing Service, Canberra, Dec. 1992.

4 Robertson, M., Vang, K. and Brown, A.J., 'Wilderness in Australia; Issues and Options', Discussion Paper for the Minister for the Arts, Sport, Environment & Territories, Australian Heritage Commission, Canberra, Jan. 1992.

5 Sinclair, J., 'Analysis of wilderness policies and legislation and preparation of options for Commonwealth Government action', Report to the Department of Arts, Heritage and Environment, Canberra, 1987.

SECTION 2

Habitats and wilderness—use without abuse

2.1

Making wildlife habitats more valuable

A CASE HISTORY FROM WILDLIFE MANAGEMENT INTERNATIONAL

The concept of 'Conservation through Sustainable Use' (CSU), as applied to wildlife today, is based on an economic reality. According to Dr Grahame Webb it is 'value-driven' conservation: by using wildlife (consumptively or non-consumptively) to generate economic benefits it is possible to create economic incentives for conserving wildlife and wildlife habitats. If the rates of use do not exceed a species' capacity to renew itself, the use—and the incentive to conserve—can continue indefinitely, and can therefore be considered sustainable. In this case history, Grahame Webb describes how CSU concepts are integrated.

I believe that to use any item continually is to use it sustainably, regardless of the form of use. There is much substance in the statement that 'conservation is sustainable use'. If the item is a renewable resource, such as wildlife, then sustainable use implies that whatever is used is replaced by the population—hence the population will be conserved.

Several aspects to sustainability, as it relates to wildlife conservation, are often ignored. For example, there can never be a guarantee that any use will be sustainable in the long term, because it is impossible to predict what tomorrow may bring. There can only be a probability of sustainable use, and sustainable use can only be measured if use is occurring. For this reason monitoring and adaptive management are important. It is also impossible to predict accurately how a population will respond to being used consumptively. Experimental use—management by experiment—is the most accurate and cost-effective way of fast-tracking conservation through sustainable use.

If plants and animals have no use, they have no value, hence they become useless or valueless. People rarely expend resources conserving them in the face of competing economic reasons to use wildlife environments for other purposes. This is especially so in developing countries, where poverty is widespread. Sustainable use can never be guaranteed, so CSU programs must be committed to monitoring, and must involve feedback so that management can be continually refined and adapted.

The advantage of using wildlife commercially is that with careful planning, private companies can pursue conservation for profit. This concept may require some serious soul-searching, but it is being experimented with all over the world. Somehow we must decide the extent to which conservation philosophy will constrain the pragmatic achievement of conservation goals.

The question begs to be asked: 'Should we buy and use wildlife products?' I believe the answer is yes, but with one simple proviso. We need to make sure that they have been derived from a successful wildlife management program—ideally, one that puts value on natural habitats. You are helping conservation by buying the ensuing products; the reverse is also true. If we choose not to purchase these products, then the lands that generate them can surely be turned over to conventional agriculture of one form or another, which will typically involve productivity and biodiversity costs. What is the incentive we create for habitat conservation each time we buy a pair of shoes made from cattle leather?

NORTHERN TERRITORY CROCODILES: AN EXAMPLE OF CONSERVATION USE

A CSU case history in action is the commercial raising of crocodiles in the Northern Territory. In this case crocodile conservation has been separated from the issue of whether people like or dislike crocodiles. By making crocodiles valuable, in terms of dollars, most of the community will see them and their habitats as assets—something 'worth' saving.

In 1973 I became involved with research on crocodiles in the Northern Territory. Two years earlier, the saltwater crocodiles had been protected after 26 years of intense commercial hunting. Adult crocodiles were rarely seen anywhere in the wild, and I was genuinely concerned about whether some critical extinction threshold had been reached. During the 1970s I, together with fellow environmentalists, studied various aspects of crocodile biology, with simple preservation as the ultimate goal. But by the end of the 1970s it was becoming clear to us that the numbers were expanding, to such an extent that 'anti-social behaviour' towards people posed the biggest threat to the public supporting crocodile conservation.

One of the critical lessons we learned was that it was almost impossible for people living thousands of kilometres away to assess the local situation realistically. We were faced with the absurd situation of people in Sydney, Melbourne and Canberra insisting that crocodiles in the Northern Territory were not recovering and were still endangered. The Federal Government was convinced that this was the case, and made policies based on these perceptions. In reality, saltwater crocodiles seemed to be everywhere and the management problem had long since ceased to be crocodile 'preservation'. Within a few months there were two fatalities, two serious maulings and a particular 17 ft (5.2m) crocodile, named 'Sweetheart', started tipping over fishing boats.

Amidst bitter criticism from environmentalists, both nationally and internationally, we embarked on an innovative management program which incorporated three critical elements. The first was education: it was vital to educate the people who lived in the north about the recovery of the crocodile numbers and how to avoid attacks. The second issue centred on what to do with 'problem crocodiles': this was accomplished by physically removing them from areas with high human habitation, such as Darwin Harbour and Gove. The third was the decision to commercialise crocodiles through tourism and crocodile farming.

Crocodile farming was initially based on captive breeding principles (using captured problem animals as breeding stock), but the real conservation value came with the 'ranching' program. This now consists of collecting wild eggs, which are then artificially incubated to produce hatchlings, and these are raised on farms until culled for skins and meat. Landowners get paid for each egg collected. With the benefit of hindsight, the program has worked extremely well. Next to Ayers Rock and Kakadu, crocodiles are an easily-recognised symbol of the Northern Territory, attracting both national and international attention. Today, people in the Northern Territory are well-informed about crocodiles, and although the wild population continues to expand, attacks occur rarely. Six farms, all privately owned and financed, are now exporting crocodile skins. The public acknowledges and accepts crocodiles as a commercial asset.

2.2

Healesville Sanctuary—recreating balance in an intensely used environment

A CASE HISTORY FROM WESTERN MINING CORPORATION

For over 60 years Healesville Sanctuary in Victoria has been a superb oasis of bushland habitat for viewing Australia's native flora and fauna. The exhibits are enjoyed by over 300 000 people annually. In recent years the sanctuary's special aim has been to display native fauna in as natural setting as possible. It has also provided a special education focus which each year allows 35 000 students to learn about the environment in what is essentially an open-air classroom.

Ecotourism takes its toll. In 1988 we realised that this very special natural refuge was under serious threat from complex problems created by an intensely used habitat, with major areas of vegetation in decline. Some of the problems resulted from natural occurrences while others arose from the lack of long-term habitat management in an intensely used area. Unless these problems were quickly dealt with, the ecosystem would be destroyed along with much of the flora and fauna that the area supported.

Much of the tree canopy was past maturity and declining rapidly. Many significant old trees had already been removed and others were dying and had to be removed for safety reasons. Natural regeneration was virtually impossible due to prevailing land conditions. The imbalance had been exacerbated by serious damage inflicted by fungal infections and insects over the previous few years. Such attacks had contributed markedly to the dieback of trees.

The accumulated land management problems consisted of a lack of resources that made long-term, comprehensive strategies impossible. As a result there was a serious decline and deterioration in drainage patterns, a lack of adequate windbreaks, the compaction and deoxygenation of soil, and weed infestation. There had been difficulties with horticultural practices and lessons were learned the hard way with soil buildup around tree bases, the over-grazing of trees and shrubs by native animals, and too much human interference by way of paths and exhibits placed too close to trees and vegetation.

Although the problems were at a critical stage, our investigations established that the situation could be reversed with a major revegetation program combined with a detailed program for protecting and rehabilitating the natural remnant vegetation. Like all major projects, a great deal of money was needed over an extended period. The Zoological Board of Victoria approached us for help. We offered not only substantial financial resources, but also technical resources.

WMC is one of Australia's largest public companies with a business focus on the discovery, development, production, processing and marketing of minerals, metals, petroleum and fertilisers. It is also one of Australia's major domestic gold producers, the world's third-largest nickel producer, and a copper and uranium producer of international significance. Our product diversity extends to petroleum and talc as well as alumina and aluminium production through a 48.25 per cent interest in Alcoa of Australia. Apart from using the profits of mining for what was clearly an important community cause, we were able to offer our considerable experience in dealing with land management issues. This included the rehabilitation of disturbed areas which is an important requisite of post-mining care for the environment.

Revegetation involves much more than planting. It re-establishes the entire ecosystem by constructing the correct strata of canopy with the appropriate species and managing the influencing factors to create and maintain the necessary balance for the ecosystem to unfold. Healesville Sanctuary provides a rare opportunity to experience five of the major flora community types in a relatively small, contained area. These five communities are:

- the riparian, or creekside vegetation—a strongly layered community with an over-storey of eucalypts, open middle stratum of *Acacia melanoxylon*, a low fern/shrub layer and a ground stratum dominated by a large range of herbs and grasses;
- the open flood plain which borders the riparian zone;
- the dry sclerophyll—a dry ridge with a canopy dominated by *Eucalyptus radiata*, a middle stratum represented by varieties of peas, wattles and heaths, and a lower stratum dominated by grasses;
- the wet sclerophyll which is similar to the dry sclerophyll but with *Eucalyptus obliqua* as the major canopy species and a ground stratum of grasses and sedges; and
- the swamp with a dominant ground stratum of grasses, sedges and herbs.

The revitalisation of these communities in the Sanctuary has occurred in three main ways: the Revegetation Project, Dieback Control and Land Management

Practices. For the revegetation program we planted indigenous species extensively to restore the diversity and structure of the plant communities. The revegetation sites were treated differently, according to the community type, remnant species and the amount of disturbance in an area. Since the project began, 100 former species have been reintroduced into the Sanctuary.

Exotic species and weeds are controlled by herbicides, although these are less widely used today because of the amount of natural regeneration throughout the areas. In some cases indigenous plants are assisting with weed control—for example, the herbaceous species commonly called fireweed is a primary coloniser and has been allowed to seed freely and grow in many areas. It has also helped to knock out exotic weeds and grasses and has allowed other indigenous grasses and herbs to regenerate. The moths and insects that use fireweed as both a food and habitat plant are, in turn, vital links in the food chain of the ecosystem.

The dieback control has involved re-establishing areas with massed planting, to allow the natural selection of disease resistance in the new plant communities and the monitoring of infestations of fungal diseases and pest insects. A research project on dieback caused by the fungus *Phytophthora cinnamomi* was conducted by La Trobe University and Alcoa of Australia. The effects of temperature, microbial populations, water, organic matter and fertility levels in the soil gave us an understanding of the fungus and have made long-term prevention strategies possible. Vegetation management is also an integral part of our strategy for controlling free-ranging animal populations to minimise damage to vegetation, weeds, and creekside erosion. A drainage study identified problem areas throughout the Sanctuary. The rehabilitation program includes the construction of a spinal drainage system for problem areas and future revegetation sites.

Other studies have included the radio-tracking of Brushtail possums in conjunction with the University of Melbourne. The information on movement patterns and foraging behaviour will help us to protect new plantings and may have wider applications for managing possums elsewhere. Low-voltage electric fencing has also been trialled in attempts to exclude kangaroos from newly planted areas.

An integral part of the land management strategy was the development of wind-break protection plantings and the restoration of vegetation diversity. In the first two years 20 000 trees and shrubs were planted to help restore a balanced structure. The canopy and middle strata species planted over this period became the 'nurse' species for the ground flora and low strata species that we have introduced since then. The success rate of the plantings has averaged around 72 per cent, varying from as high as 90 per cent to as low as 44 per cent. Difficulties are caused by various factors including severely disturbed soil profiles, unsuitable climatic conditions at planting times, poor weed management and grazing by macropods and Brushtail possums.

The enrichment of creekside areas by planting and bank stabilisation has enhanced Badger Creek and the creekside community. The numbers of free-flying parrots, wrens and other birds using the revitalised bushland habitats have already increased.

In addition to the public exhibit area, the Sanctuary has managed the last unused remnant of the former Coranderrk Reserve since 1955. This encompasses 142 hectares and is one of the few remaining areas of forest in the Yarra Valley and an invaluable resource for our understanding of the natural habitat. In an increasingly altered environment, the Coranderrk and the Sanctuary bushland provide an 'island' of refuge for animals and a much-needed vegetation corridor of trees along Badger Creek. The Sanctuary maintains special breeding programs for various endangered species such as the Leadbeater's possum, the Long-footed potoroo and the Eastern barred bandicoot.

Our Revegetation Project at Healesville Sanctuary is based upon the restoration of habitat; this project benefits not only the diverse number of animal species found in and around the area, but also helps us to educate the visitors. Healesville Sanctuary is a unique bushland experience. Surrounded by farm land and housing estates, the Sanctuary acts as a genetic pool for both flora and fauna. It provides the remnant corridors that are critical for the future survival of many local species. Our current experimental maintenance and monitoring work will provide information not only for long-term management plans for the Sanctuary and its connecting bushland areas, but also for ecotourism projects and other regeneration projects—in Australia and worldwide.

2.3

The facts, fictions and frustrations of the forest industry

A case history from the National Association of Forest Industries

The National Association of Forest Industries Ltd (NAFI) represents companies, individuals and organisations involved in the forest and forest products industries throughout Australia. Its activities are funded entirely by its members, who between them represent all facets of the industry and produce most of Australia's forest output.

NAFI represents the forest industry's views to governments and public authorities on matters relating to the development and use of forests and forest products. It also promotes community awareness of the forest industry by providing information and educational material.

The forest industry in Australia employs about 85 000 people in forest management, timber harvesting and downstream manufacturing. This makes it about the same size as the motor vehicle and parts industry, and bigger than the clothing and footwear industries, and the coal-petrochemical industry.[1] More important, with this workforce, twenty paper mills, and some 1600 sawmills, the industry still does not produce enough to satisfy Australia's need for the final products. Annually some $2.4 billion worth of product is still imported.[2]

There is considerable growth potential in the forest industry. The Industry Commission's September 1993 Report used a model to estimate that industry expansion could be worth an additional 0.3 per cent or about $1.2 billion per year in real GDP.[3] However, the road to economic growth in the forest industry is fraught with environmental conflict. Conservation groups warn of the destruction of rainforest and ecosystems, and battle with loggers who are equally concerned with sustaining the resource, but for different reasons, including maintaining their livelihoods. This situation is highlighted from time to time in the media and has led to much misunderstanding and misinformation.

Before European settlement less than 10 per cent of Australia was covered by forest. Since then, over one-third of the forests have been cleared for farms and urban development. Even so, Australians still have more forest per capita than 90 per cent of the world's countries.[4] Australia has 43 million hectares of native forest, or about 5 per cent of the continent. About 10 million hectares of native forest are in national parks and other conservation reserves. These areas are permanently reserved from logging.[5]

Of Australia's native forests, 11.5 million hectares are state forests managed for multiple uses including sustainable timber production. Some 7.1 million hectares are specifically managed for perpetual wood supply (managed for wood production on a sustainable yield basis and to conserve biological values). Only about 1 per cent of this is harvested in any one year. The rest is either inaccessible, uneconomical, or set aside for environment protection purposes under various Codes of Forest Practice.[6]

Some 60 per cent of Australia's native forest estate is regrowth. It has either regenerated naturally after fire, such as in the Central Highlands of Victoria and much of south-east New South Wales, or has regrown after harvesting. The quality of many regrowth forests is so high that some areas which have been harvested by the industry for more than 100 years are in national parks and World Heritage areas. These include parts of the south-east forests of New South Wales; the Daintree, and Fraser Island in Queensland; and much of Victoria's alpine ash forests.[7]

The main forest types in Australia include rainforests and mangroves (8 per cent), eucalypts (60 per cent) and the native pine and open forests (32 per cent). Eucalypts are the main timber production resource. Of all the forest types, about 40 per cent are unlogged.[8]

In recent years the forestry industry has had to endure harsh treatment and criticism as forestry is a very emotional media and community issue. For example, no-one has a problem frying a 50-year-old sea perch, yet these fish have about the same growth cycle as an average tree. There seems to be an acceptance of 'ocean farming', whereas the harvesting of trees becomes a passionate media issue. According to Dr Robert Bain, Executive Director of the National Association of Forest Industries, the forest industry farms many trees—like any other crop—and selectively harvests them at the optimum point for a maximum quality product. He says:

> The forest industry is not out to destroy forests. We have more real interest in making sure the resource we depend upon is not depleted than any conservation group could possibly have. That includes old-growth forests. There has been a vast number of reports clearly acknowledging that responsible harvesting has no major negative effect on forest ecosystems, yet we still are plagued by the emotional attacks of conservation groups, and political indecision and back-pedalling.

The multiplicity of issues and opinions surrounding this controversy seems to defy resolution. Under the title 'Managing Conflict' the Ecologically Sustainable Development (ESD) Working Group's Final Report on Forest Use of November 1991 made these comments: 'With the increased focus of the community on environmental matters, decisions on forest use have generally been taken in a climate of intense political pressure and public debate'.[9]

The ESD Working Group commented that approaches to resolving value conflicts 'have tended to be ad hoc and address single point conflicts with short-term solutions rather than develop broad regional forest-use strategies with ecologically sustainable development objectives'.[10]

To give an idea of the amount of material generated by government, six major reports on Australia's forest industry were produced in three years: the National Plantation Advisory Committee's Integrating Forestry and Farming (November 1991); the final report on forest use by the Ecologically Sustainable Development Working Group (November 1991); the Resource Assessment Commission (RAC) Forest and Timber Inquiry, Final Report, March 1992; the National Strategy for Ecologically Sustainable Development (December 1992); the National Forest Policy Statement (December 1992); and the Industry Commission's Final Report Adding Further Value to Australia's Forest Products (September 1993). The RAC Forest and Timber Inquiry was chaired by Justice Stewart, received 550 written submissions and took 4400 pages of transcript from 423 witnesses in 50 days of public hearings during its period of two years and four months.[11]

Unfortunately, all these initiatives have contributed more branches to the tree of knowledge without resolving the politics. All the reports have tackled the question of how sustainable development relates to the forest industry, but there is not even any real consensus on the definition of sustainable development, nor on how it affects the forest industry. For example, the World Commission on Environment and Development (the Brundtland Commission) set the broad parameters: '. . . development that meets the needs of the present without compromising the ability of future generations to meet their own needs'.[12]

In relation to the forest industry, the United Nations Conference in Rio de Janeiro in June 1992 concluded: 'Forest resources and forest lands should be sustainably managed to meet the social, economic, ecological, cultural and spiritual human needs of present and future generations'.[13]

For Australia, the Resource Assessment Commission recognised the difficulties: 'Some inquiry participants emphasised economic issues, others emphasised ecological issues and others placed emphasis on social issues. In short, the concept means different things to different people'.[14]

The ESD Working Group also addresses these issues:

> The maximisation of community benefits from forests necessarily involves an assessment of the competing values of the range of forest uses and the allocation of property rights to the highest value mix of uses. That may mean

> conservation of biodiversity in some cases, and in other cases, of multiple use including integrated wood production operations producing sawlogs and woodchips.[15]

The ESD Working Group also created its own simple definition under the heading, 'The Working Group's view of ecologically sustainable forest use':

> The principles of ecologically sustainable forest use will require the development of a policy framework and of approaches which recognise the following three requirements: maintaining ecological processes within the forests; maintaining biodiversity; optimising benefits to the community from all uses within ecological constraints . . . This involves optimising commercial benefits within ecological constraints; optimising intangible values and non-commercial uses consistent with achieving optimum net benefits; and the maintenance of options for future generations.[16]

A table on sustainability and forestry used in 'Caring for the Earth—a strategy for sustainable living' classifies ecosystem conditions. The potential for sustainability ranges through (1) natural systems, (2) modified systems, (3) cultivated systems and (4) built systems. Degraded systems are classified unsustainable.[17]

In forestry terms, the key conditions are: Natural Systems—described as 'ecosystems where, since the industrial revolution (1750), the human species impact has (a) been no greater than that of any other native species, and (b) has not affected the ecosystem's structure';[18] Modified Systems—described as 'ecosystems where human impact is greater than that of other species but whose structural components are not cultivated. Most of the planet is now modified including land and sea areas usually considered "natural". For example, naturally regenerating forest used for timber production and naturally regenerating range land used for livestock production';[19] and Cultivated Systems—described as 'ecosystems where human impact is greater than that of any other species and most of whose structural components are cultivated: e.g. farm land, sown pasture, plantations, aquaculture ponds'.[20]

In this context, most natural forests now fit into the 'modified forest' category meaning 'a forest where trees have been felled (during the past 250 years) usually by loggers or shifting cultivators or where other products have been harvested but where cover of trees or shrubs of indigenous species persists'.[21]

The Resource Assessment Commission has modified the IUCN table and uses the same concept starting with unlogged forest and moving through regrowth forest, plantations, crop land and pasture, and degraded land systems:

1 Unlogged forest: also known as primary, pristine or undisturbed forest. Unlogged forest includes old-growth forest.
2 Regrowth forest: results from logging or pastoral grazing but where indigenous tree and shrub cover is retained. New tree growth may be

derived from natural regeneration or supplemented by specific treatments.

3 Plantations: usually former native forested land in which most trees have purposely been planted and includes native and exotic species. In many cases, the pre-existing native forests were removed originally to make way for another land use, for example agriculture.

4 Crop land and pasture: sometimes former native forested land, now under agriculture, including agro-forestry and orchards.[22]

Convoluted definitions and complex politics aside, the question that remains for the general public is whether or not Australian forests are being sufficiently protected. The RAC has concluded that 'the cessation of wood production activities in native forests is not justified on the basis of evidence before it'.[23] It added: 'The Inquiry is satisfied that currently the agencies have in place sustained yield management strategies for wood production. The evidence before the Inquiry is that these strategies are appropriate'.[24]

On species extinction, the RAC found 'no evidence to suggest that the risks of extinction resulting from logging present an immediate threat to the ecological processes on which forest systems depend'.[25] Their conclusion supported the findings of the ESD Working Group's Final Report on Forest Use.

In comparing forestry with agriculture, the ESD Working Group proposed that 'By any conceivable yardstick, forest use and management compares favourably with the environmental impacts of the agricultural sector'.[26] The Working Group also recognised the industry's record concerning species. 'There have been no known extinctions of Australian native fauna and flora species as a result of timber harvesting operations over a period of more than 100 years'.[27]

In discussing harvesting, the ESD Working Group noted:

> Forests are generally robust ecosystems and are capable of surviving short-term perturbations. Human disturbances may result in the destruction of individual animals, trees and plants, but this need not threaten the forest ecosystem as the forest has the capacity to reconstruct, to regenerate habitat and to maintain genetic diversity.[28] . . . There have been significant changes in forest management over the past two decades which are in keeping with the goals of ecologically sustainable development. These include: increased environmental regulation of forest harvesting activities; increased allocations of forest to reserves; increased research activity; an improved knowledge of the forest ecosystem; structural changes that have improved industry efficiency; and a greater emphasis on valuing all forest uses in land-use decisions.[29]

A few definitions and a couple of case histories will, hopefully, put a perspective on the efforts of some forest industry companies. The primary products of the forest industry are sawlogs and veneer timber. What remains goes into making woodchips. The sources for woodchips include: native pulp-logs which are weak, diseased or defective trees removed in the harvesting process to ensure proper conditions for regeneration; plantation pulp-logs—the thinnings (trees that are small and must be thinned out); the harvest of short-rotation pulp-log crops; and forest residues such as branches, and sawmill residues of sawlog offcuts and waste.[30]

The logs used to make woodchips are called pulp-logs because the woodchips are processed into pulp for paper products. Eucalypt woodchips make pulp that gives paper certain qualities such as smoothness and the ability to hold ink on its surface. Paper made from pine pulp is particularly suitable for newsprint, tissue and cardboard.

Woodchips are a product of integrated harvesting. The Resource Assessment Commission in its final report noted that woodchips create an important market for the residual wood in many forests, that is, 'the trees that are unsuitable for sawlogs and that would be left standing in the forest, so leading to poor stand structure, or be cut down and left to rot or burn'.[31]

Generally less than half the wood harvested in a native forest is suitable for sawlogs. Integrated forestry does not mean every tree is harvested. Codes of Forest Practice ensure that within each harvested area, habitat trees, streamside buffers and wildlife corridors are retained as habitat for the forest fauna. Regenerated areas can then be recolonised.

In some regions integrated harvesting has become standard practice, particularly since the development of the export woodchip industry in 1970. It has given the forest services the opportunity to market wood that would be wasted otherwise.[32] The use of saw-log residues as woodchips, or other residues which are made into fuel, particleboard and other products, is a competitive multi-million dollar business. However, like recycling, each opportunity must be assessed individually since several factors determine whether or not a woodchipping operation is cost-effective.

Much of Australia's forestry industry is farmed. The most significant feature of the timber products industry over the last fifteen years has been a decline in sawn hardwood and a corresponding increase in sawn softwood. These trends seem to reflect resource availability (a decrease in hardwood logs and an increase in plantation grown softwood) along with more aggressive marketing by softwood producers. Australia now has 915 000 hectares of pine plantations yielding some 5 million cubic metres of timber annually.

There are also about 85 000 hectares of eucalypt plantations or tree farms. This is increasing quickly. One company, North Forest Products, owns and operates Australia's largest eucalypt tree farm program in northern Tasmania and this plantation is growing by more than 3500 hectares each year. At Bunnings in Western Australia there are plans to develop eucalypt tree farms up to the rate of about 4500 hectares a year as company projects, joint ventures or as plantings for other investors. Most of this is suitable for wood production forestry and up to 80 000 hectares is suitable for intensive tree farming.

The research and technology base for North Forest Products tree farm program in north-west Tasmania is the Eucalypt Research Centre, a world leader in research into the breeding of cool-climate eucalypts. Research is directed at maximising the genetic quality of eucalypt plantings by breeding and cloning. In addition, the Eucalypt Research Centre has direct links with the CSIRO Division of Forestry and Forest Products, the University of Tasmania, various other Australian state and private research bodies, and overseas research institutions.

North Forest Products is growing genetically improved Tasmanian Blue Gum *(Eucalyptus globulus)* and Shining Gum *(Eucalyptus nitens)* from the North research program at its own seed orchards, which now supply all North's requirements for tree farm establishment. Besides tree farm technology, North applies sophisticated stand and land management technology such as the GIS (Geographic Information System) which stores, analyses, retrieves and presents data, and the GPS (Global Positioning System) which improves the efficiency and accuracy of data collection in the field. Incorporating genetically improved stock with other research-based management improvements, and adding surrounding private property through joint venture arrangements make it reasonable to anticipate sustainable yields of about 2 million tonnes from north and north-west Tasmania in the future.

Another major player in the pulp and paper industry is Australian Paper (formerly APM). This Amcor subsidiary has created the country's largest private forest plantation in Victoria's Gippsland region by planting 47 million pines and 14 million eucalypts in an area which was largely degraded farm land. Working closely with the CSIRO Division of Forest Research made it clear that a mix of softwood and hardwood was appropriate. Subsequent breeding programs have proved that *Pinus radiata* and three eucalypt varieties are particularly suited to the area. These 'super trees' are carefully monitored from nursery to harvest. The pine trees in particular yield up to 60 per cent more than earlier generations did. A popular misconception is that pine trees degrade the soil and that it is hard to grow other trees or plants afterwards, but stands of eucalypt planted after pine crops have been harvested have grown well. What is more, Australian Paper Forests has recently supplied 750 cubic metres of pine forest soil for the gorilla enclosure in the Melbourne Zoo because of its suitability for rainforest vegetation.

The planning, technology and collective knowledge of the forest industry in Australia is substantial and will allow both large and small companies to expand and compete internationally. Companies, individually and collectively, have also taken other measures to improve their competitive position. Individually they have increased their efficiency with investment and workplace reforms; collectively, they have worked on developing markets and improving the industry's public and political profile.

Governments have tended to lag in establishing reforms. Much government attention has focused on how the industry deals with environmental issues or on its land use decisions. Unfortunately, this has reinforced the impression that the forest sector is an environmental issue rather than an opportunity for sustainable industrial development. Of course, any industrial activity is contingent on public approval. Industry has responded to public concerns by overcoming the environmental effects of timber harvesting and production effluents. It has also greatly increased public awareness and designed an education program, and there is now evidence that the public is more prepared to understand and accept the forest-based industries.

Notes

1 National Association of Forest Industries (NAFI) submission to Caucus Employment Task Force, September 1993, p. 1.
2 ABARE, *Agriculture and Resources Quarterly*, 4(3), September 1992, p. 451, AGPS Canberra.
3 Industry Commission Draft Report, 'Adding Value to Australia's Forest Products', May 1993, p. 262.
4 Ewing, A.J. and Chalk, R. 'The Forest Industries Sector: an operational strategy for developing countries', *World Bank Technical Paper No. 83, Industry and Energy Series*, 1988, p. 70, Washington DC.
5 Resource Assessment Commission, *Forest and Timber Inquiry Final Report*, 1992, p. 2, AGPS Canberra.
6 National Association of Forest Industries, *Forest Facts*, Dec. 1992.
7 Ibid.
8 Ibid.
9 Ecologically Sustainable Development (ESD) Working Group, *Final Report*, 'Forest Use', Nov. 1991, p. 23.
10 Ibid.
11 Speech by Justice Stewart, Chairperson Resource Assessment Commission, to a Conference on Australia's Timber and Forest Industry, 'A Strategy for the Future', 28 May, 1992.
12 World Commission on Environment and Development, *Our Common Future*, 1987, Oxford, Melbourne, p. 8.
13 United Nations Conference on Environment and Development, June 3–14, 1992, Agenda 21, Rio Declaration, Forest Principles, Principles and Elements, p. 3.
14 NFI, 'The National Forest Inventory Objectives', *National Forest Inventory News*, vol. 1, 1991, p. 3.
15 ESD, op. cit., p. 33.
16 ESD, op. cit., p. 35.
17 *Caring For The Earth: A Strategy For Sustainable Living*, Oct. 1991, published in partnership by IUCN—The World Conservation Union, UNEP—United Nations Environment Programme and WWF—World Wide Fund for Nature, Switzerland, p. 126.
18 Ibid, p. 126.
19 Ibid, p. 126.
20 Ibid, p. 126.
21 Ibid, p. 123.
22 RAC, p. 108.
23 Ibid.
24 Ibid, p. xxxix.
25 ESD, op. cit., p. 163.
26 ESD, op cit., p. 29.

27 Ibid.
28 ESD, op. cit., p. 44.
29 ESD, op. cit., p. 29.
30 Resource Assessment Commission, Forest and Timber Inquiry, *Final Report*, 1992, vol. 1, p. 37, AGPS, Canberra.
31 Ibid, p. 37.
32 Ibid, p. 302.

2.4

Turning environmental problems into economic opportunities

WINSOME MCCAUGHEY, CHIEF EXECUTIVE OF GREENING AUSTRALIA

Ten years ago only a handful of local communities, mainly in Victoria and Western Australia, were concerned with land degradation. Now there are over 2000 rural community groups and at least another 600 urban community groups committed to tackling problems and restoring the soil, water and vegetation in their localities. Many of these groups work closely with other neighbouring groups, local councils and government agencies in their state. Winsome McCaughey, CEO of Greening Australia, explains how Landcare works.

'Landcare' is a generic term covering the hundreds of rural groups that have formed over the last decade, under many different names, to tackle many different problems: for example, farm tree groups, weed eradication groups, rabbit control groups, and soil protection groups. 'Landcare' is also the term used to describe Commonwealth and state government funding programs geared to help these groups.

'Greening' refers to restoring vegetation to sustainable land management levels—an activity most groups undertake. Greening is also part of the name of the community-based Greening Australia Association; many of these rural and urban groups are members, as are local councils and individuals. Greening Australia was started in 1982 by the United Nations Association of Australia and the Nursery Industry Association of Australia. We manage the Federal Government's One Billion Trees Program and the National Corridors of Green Program under contract, and we educate and advise the community on revegetation projects and local greening plans that combine economic and environmental benefits.

The landcare and greening movements are being led by people who do not fit the usual public perceptions of what constitutes a 'greenie'. The leadership has come mainly from the farming communities—from the people whose families have been working the land, sometimes for generations, but who are now gradually realising that there will be no economic future if the environmental issues associated with land management are not speedily resolved. It is this part of the farming community that is raising the awareness and empathy of the general public and alerting them to the fact that our future lies in making our land, water and vegetation management practices much more sustainable. Without this, our soil quality—that most precious base—will be gone.

Although Australia is the driest continent on Earth (less than 10 per cent of its land is arable and less than 10 per cent had forest cover) the early European settlers saw it as a country of limitless wealth and endless rolling pastures. Unfortunately, our native forests, bush and grasslands have been cleared to make way for urban settlement, agriculture, forestry, mining and many other purposes. This clearing, combined with unsustainable land management practices, has led to a massive set of interrelated environmental problems; these threaten not only biodiversity and the complex web of life on which we all depend, but also our economy. Our agricultural export opportunities still account for a significant part of our export earnings, and it is only the health of our soil and water that makes this possible.

However, some two-thirds of the continent is degraded and in urgent need of land repair. Our soil is compacted from hard-hooved animals, and it is literally washing and blowing away. Wind erosion is not a new problem; in 1902 the Albermarle station of over 40 000 hectares near Wilcannia in New South Wales reported losing soil to a depth of up to 30 centimetres at a time from the huge dust storms which blew across their drought-stricken land. The dust storms that raked the continent in May 1994 were conservatively estimated to have cost Australia $100 000 000 in lost soils. Dr John Williams, the Deputy Chief of CSIRO's Division of Soils said that the 'mass clearing of our native vegetation was largely responsible for the topsoil losses by wind erosion'. Since it takes up to 200 years for a few millimetres of soil to form, we cannot afford to allow this irreplaceable and invaluable resource to be simply blown out to sea. Replanting the vegetation is the most cost-effective form of prevention available. Recent estimates of soil erosion in the Darling Downs in Southern Queensland indicate that the region is losing about 8 million tonnes of the rich black topsoil through erosion every year.

The salinisation of soil and water is a massive exponential problem for Australia. In Western Australia alone there are 16 million hectares of cleared agricultural land. The Western Australian Department of Agriculture estimates that about 2 million hectares of these will go out of production over the next few years unless major revegetation is not established to lower saline watertables and bind the earth's surface. In February 1994 Western Australia's Government launched a $60 000 000 joint venture with Oji, Japan's largest paper firm, to establish 20 000 hectares of trees on farms, so creating a new wood reserve worth up to $1 billion a year. The state's farmers intend to use the scheme to tackle salinity and diversify income.

Australians are becoming very familiar with the 'wetland' salinity problems caused by irrigation, but the lesser known 'dry-land' salinity is an ecological time-bomb for much of our continent. Geologically, Australia is an ancient land with substantial areas once covered by the sea or salty lakes millions of years ago. The retreat of more salty waters left high concentrations of natural salts underground. Before Europeans began extensively clearing the land about 200 years ago, the ecological system in our arable lands was in balance and the watertables were well below ground. But the removal of the deep-rooted vegetation and inappropriate farming practices have led to these underground natural salts being pushed to the surface, and these rising levels of saline water render the soil virtually useless.

Salinity is also ruining many of our water systems including the Murray network in which every minute about 2.5 tonnes of salt flow across the South Australian border. Recent studies suggest that Australia will have to spend about $11 billion dollars between now and the year 2000 on restoring and maintaining our water quality. The cost of building a new reservoir to replace a major urban water supply degraded by salt can vary from $50 to $100 million alone.

The vegetative cover is under threat not only from current human practices but also from our legacy of feral weeds and animals. The invasion of 'woody weeds' in parts of the far-west of New South Wales and Queensland and along much of our waterways is strangling much of the remaining indigenous vegetation. Similarly rabbits, wild horses, pigs, goats, buffaloes and donkeys are destroying large areas covered by fragile vegetation across various parts of the continent. The indigenous Australian fauna that depend on the indigenous vegetation cover for their survival are also under great stress. Some 27 species of mammals and birds are now extinct and a further 59 species of mammals, birds, reptiles, frogs and fish, plus hundreds of invertebrates, are on the endangered species list.

However, all is not bleak as the rural part of the landcare/greening movements now involves about 39 000 people. Perhaps the best way of describing the scale of the problems facing these groups, and the depth and breadth of their response, is to take a closer look at one of Australia's very first Landcare groups, the Warrenbayne Boho Land Protection Group.

This group lies in the north-east of Victoria, in the foothills of the Strathbogie ranges, where my family has lived since soon after European settlement. The group evolved in 1983 after a few landholders in the district began sharing their concerns about the rate at which the saline underground watertables were rising, and the extent of land that was under threat and going out of production. They realised that if there was any hope of combating the area's salinity and other land degradation problems across the area, they and their neighbours would have to shoulder the responsibility for the problems and solutions.

The group now includes 150 landholders across an area of about 40 000 hectares. They estimate that some 11 000 hectares of this land needs to be revegetated with trees and/or other deep-rooted perennials if their own land and the land and water quality further down the catchment is to be saved. The group realised that the problems they faced were well beyond their financial or technical resources, so a decade ago they set about engaging the support of government agencies, tertiary institutions, corporate and philanthropic bodies, urban people and anyone who would assist them.

In the mid-1980s the Commonwealth Government's National Soil Conservation Program (now the National Landcare Program) funded the group to conduct a landholder-based extension project. This enabled the group to appoint a full-time co-ordinator from their own farming community to conduct their own training and research and to help landholders develop sustainable land management plans for their own farms along with those of their neighbours.

Ten years later the group estimated that nearly 1000 of the 11 000 hectares needing revegetation are back under vegetative cover, with large areas of land still in need of revegetation. The group then considered how to tackle the conservation problems with commercial solutions, since the only conceivable way to restore vegetation cover was to turn the environmental problems into economic opportunities.

The group is now encouraging not only its own landholders, but also those in the neighbouring landcare groups which have grown up around them, to incorporate timber belts, shelter belts and farm woodlots into their farm landscapes for a combination of: land protection (for lowering saline watertables and prevention of erosion); animal and stock protection (shade from sun and shelter from winds); and commercial wood return (by providing a steady flow of high-quality wood on a scale suitable for regional timber industries and wood processors).

It is ironic that this country has cleared at least 20 billion trees (to the great destruction of the land), must re-plant many billions of trees (if it is to restore and maintain what is left of our economic land base and our ecosystems), and yet continues to import around $2 billion worth of timber and associated wood products every year. If we were to grow this wood ourselves in strategic plantings, integrated into up to 20 per cent of individual farms, we could be helping to save our precious soil, diversifying farm incomes and reducing our foreign debt. Why do we continue to import wood rather than grow it when so many Australian species are known to be superb for both timber and pulp? Forty per cent of the world's commercial wood plantations are Australian species, but only a tiny fraction of these plantations are grown in Australia.

The conservation and commercial plantings of the Warrenbayne Boho landcare group are now being incorporated into a much broader farm forestry scheme which proposes planting 120 million native trees in the Murray–Darling Basin. The scheme is driven by the Victorian Government's Trees for Profit Program (supported by the Federal Government), and a private Victorian-based company, Tree Corp. It aims to earn $1 billion in hardwood exports and beat the nation's worst environmental scourge—salinity.

Some of the newest developments in turning around environmental problems through economic development are the Federal Government labour-market programs; these involve unemployed people in conservation and commercial wood projects. An interesting illustration of this is found on the far north coast of New South Wales (centring on Lismore) where there are severe water, soil and weed infestation problems. The region contains the remnants of what was once a magnificent sub-tropical rainforest known as the 'Big Scrub', originally covering 75 000 hectares, of which less than 500 hectares now remain in isolated fragments. The region once sustained a thriving timber industry but resources for this have steadily diminished. The area also suffers high rates of unemployment—up to 20 per cent of its population, many of whom are very skilled and committed people, are out of work.

For the last three years Greening Australia (with funds made possible by the One Billion Trees Program) has set up programs involving some 600 unemployed people in restoring the vegetative cover, partly for conservation purposes but also as a means of renewing the region's wood resources. These programs are geared to securing and restoring the remnant vegetation, as well as to developing demonstration farm–forestry projects including collecting and propagating seeds, and planting seedlings along the waterways and in the farm–forestry projects. Some of the participants also map the remnant vegetation and plot the location of future vegetation and Corridors of Green. Their involvement will also help to ensure that there is a trained workforce for future forest industries.

2.5

Ensuring that agricultural land is a long-term renewable resource

RESEARCH ISSUES FROM CSIRO

Human population growth and its consequences are the most vexatious issues in the debate on biodiversity. According to the CSIRO, which is working to make food production more efficient, Australia currently feeds perhaps 100 million people worldwide with its agricultural products. Since most of the continent has poor soils and unpredictable climate, that is an impressive achievement, but it comes at considerable environmental cost.

The CSIRO Division of Wildlife and Ecology has found that pastoral farming occupies a large proportion of all rangelands, while only 3 per cent is protected in national parks and conservation reserves. Not surprisingly, nearly half of the rangelands' original native mammals are no longer found there. The Federal Government's Working Party Report on the conservation of biodiversity, as it relates to sustainable development, identifies the destruction of natural habitat as the biggest threat to biodiversity in Australia, and urgently recommends a comprehensive system of conservation reserves for terrestrial, freshwater and marine environments. It is quick to point out, however, that since many species do not occur in areas likely to be conserved in the proposed reserves, we must develop ways to maintain biodiversity outside reserves. This will involve the control of exotic weeds and pests, the management of remnant habitats used in agriculture and timber-getting, the development of agro-forestry and alternative farming techniques, and the encouragement of community conservation programs.

However, the biodiversity report makes no mention of human population. What, for example, is the long-term carrying capacity of this country, and of the world?

How efficiently are people using nature's resources? A 1986 study of global human activities by Professor Paul Ehrlich and colleagues at Stanford University estimates that humans appropriate nearly 40 per cent of the net primary production (NPP) of the Earth's land areas—that is, the total amount of energy fixed biologically, less the energy needed by the primary organisms (chiefly plants). The NPP is a measure of the planet's total food resource.

If the estimate is correct, the implications for biodiversity look rather stark. While one species is reserving 40 per cent of the terrestrial food resource for its own purposes, the other millions of terrestrial species are being asked to survive on the remaining 60 per cent, and if the human population continues to grow at current rates, we can reasonably expect Australia's share of NPP to rise to 100 per cent before the end of the twenty-second century—an ecological impossibility.

However, the Stanford University study makes an interesting distinction between the food needed by humans to survive and the food we waste or consume indirectly. The study estimates the amount of organic material consumed directly by humans or domestic animals at only 4 per cent of NPP—just one-tenth of the energy appropriated by humans. The rest—more than a third of the total land NPP—is wasted or denied to other species by burning organic material when clearing land, discarding parts of the crop plants we harvest, and covering the land with urban developments.

BUSH TUCKER—A FUTURE STAPLE FOOD?

CSIRO believes that the fate of the planet's biological diversity may depend at least as much on how efficiently we exploit the areas not protected in reserves as on what we protect in them. One of our strategies is more efficient crop land. At CSIRO we are studying traditional Aboriginal bush tucker including Australian wattle tree seeds which could become a more widely used human food for the world. So far Australian plants really have not contributed much to the world's dinnerplate. Despite our rich heritage of native flora only one Australian plant—the macadamia nut—is grown widely as a food crop.

Aborigines in central Australia have used acacia seeds as a seasonal food for thousands of years. Recently, there has been renewed interest in bush tucker—including wattle-seed flour—in Australian gourmet circles. We believe there is also great potential to use wattle seeds to help alleviate hunger in semi-arid countries, many of which already grow Australian acacias on a large scale for firewood and as wind-breaks around crops. Chemical analysis of acacia seeds shows that they are quite nutritious as they are rich in protein, carbohydrate and fat. Already taste trials in Africa have found that over 20 local recipes incorporating acacia seed flour are highly palatable to local people.

The Australian Tree Seed Centre, part of CSIRO's Division of Forestry, is working with specialist laboratories in Australia and overseas to determine which species of acacia are most nutritious and to make sure they contain no toxic substances. The seeds must meet international standards if they are to be widely accepted as a human food. Our tests so far have been very encouraging. We have also been working with researchers in Africa to test the growth and seed production of different acacia species in different African environments. Combining traditional Aboriginal knowledge and modern scientific research in this way is an exciting prospect.

GRAZING ON GLASS

Another environmentally sustainable project aimed at ensuring stable grazing land comes in the form of a research project nicknamed GLASS, which is improving the speargrass grazing lands of northern Australia. The GLASS project (an acronym for Grazing, Legumes, Land classes and Stability on pastures in the Sub-tropics) aims at providing graziers with management options that will maintain pasture productivity and stability and improve profitability.

Speargrass country, which generates about $500 million a year for the Queensland beef industry, includes several fragile granite types of grazing land. The project is examining how stocking rates, introduced legumes and burning affect the stability of native and grass legumes pastures, in the hope of identifying options that will allow graziers to maintain profitability without destroying the land for long-term use.

The GLASS project is based at Glenwood, a commercial property 55 kilometres west of Mundubbera in Queensland. The site, made available by the McIver family, is ideal for this type of research as it has several types of granite land on it and has been grazed for over 100 years. The importance of the GLASS project is reflected in the team involvement of CSIRO and the Department of Primary Industry. It has also attracted substantial funds from the Meat Research Corporation and the National Soil Conservation Program, along with support from Incitec and Monsanto.

Early results are indicating strong links between soil and pasture condition and animal production. Stocking more than one beast to 1.7 hectares, particularly on narrow-leaf ironbark country, appears to lead to higher soil erosion, less rain going into the soil, a lower quantity and quality of pasture, and a lower animal production. However, we are getting increases in live-weight gains of 40–56 kg a head on three-year-old legume pastures at a stocking rate of one beast for every 2 hectares. We now have some preliminary ideas about how graziers could improve the efficiency of herds on speargrass pastures for a profitable and sustainable enterprise.

MORE FOOD, FEWER CHEMICALS

Without pesticides, the prospects for continuing to feed the world's rapidly growing population would be bleak. However, concern about their potential for serious environmental consequences has prompted a search for new alternatives.

Chemical residues from some insecticides persist in the ecosystem, with traces finding their way into wildlife, livestock, and people. Many insecticides are indis-

Above Healesville Sanctuary Nature Trail before it was replanted with native species—early 1990. *Below* After replanting with native species—late 1990.

Top left Logged timber for wood chipping; it is unsuitable for timber production because of its shape and small diameter. *Top right* Woodchips: an essential stage in timber and paper production. *Above* A native tree nursery: the plants are used for the regrowth of natural forests.

Above RGC Wetlands Centre, Capel, Western Australia: a wetland ecosystem being developed on lakes created when mineral sands were mined. *Below* North Stradbroke Island, Queensland: rehabilitated frontal dunes 12 years after mining. These dunes are now stable and support a range of habitats including spinifex grasslands, shrublands and wetlands.

Above The Pebble-mound mouse. Hamersly's environmental studies for its new Marandoo mine have provided reassuring information about the habits and populations of this mouse species. *Below* Zoologists installing fauna pit traps at the Marandoo mine as part of Hamersley's environmental study program.

criminate in their effects, so they destroy 'good' insects—those that feed on pests or have other important roles in the natural ecosystem—along with the 'bad'. Ideally, we should target specific pest insects with sprays that have no effect on other species and leave no hazardous residues. Scientists in our Division of Entomology are working on one possibility in bollworms—caterpillars of the genus *Helicoverpa*—which cause damage that costs Australian crop farmers about $450 million a year. Bollworms happily eat cotton, maize, sorghum, tomatoes, peanuts, and other valuable crops, destroying not just the leaves but also the seed heads and other parts. The adult moths, however, do no damage at all. Bollworms have been a target of chemical warfare for decades, with some unfortunate results. For one thing, toxic chemicals have been accumulating in the environment; for another, the pests have started to develop resistance.

A novel approach to reduce our dependence on chemical pesticides is clearly timely, and we are investigating the use of a specially fortified virus. Viruses are tiny parasites of the cells of living organisms. The virus attaches itself to the cell with molecules on its outer coat that latch onto complementary receptor molecules on the cell membrane. Once latched on, the virus can enter the cell, where its nucleic acid (carried in its protein shell) gives instructions to the cell's enzymes to produce millions of copies of viral nucleic acid and protein coats. The cell's own life processes are disrupted and usually it dies—but not before it has served its viral hijacker and manufactured more particles to infect neighbouring cells. It so happens that a group called the nuclear polyhedrosis viruses (NPV) infect *Helicoverpa* caterpillars in the wild and can indeed kill them. The younger the caterpillar, the more susceptible it is. Adult moths are not susceptible to the virus.

Caterpillars growing on plants ingest polyhedra. These dissolve in the gut, releasing the individual virus particles to infect the cells of the lining, and the cycle starts again. How long the virus takes to kill the caterpillar depends on the temperature, but at 25°C most survive for about five days after infection. Lower temperatures extend the period. During this time, the caterpillars carry on eating, so the virus naturally present in the field already does not help very much in curtailing damage to crops.

If NPVs are so slow-acting, one may well ask why the CSIRO is working on them as potential viral insecticides. We believe it is possible to use the virus as a means to get a fast-acting toxin into the caterpillars. Such a poison must affect insects only, and be completely innocuous to vertebrates—mammals, birds, reptiles, amphibians and fish—although the virus cannot reproduce in their cells. Similarly, other insects that may be susceptible to the toxin are still safe because the virus can reproduce only in closely related *Helicoverpa* species.

The toxin must be a protein, and the information directing its manufacture must be encoded in the virus' DNA. Infected insect cells, obeying the instructions carried by the virus, would manufacture it along with the other viral components. They would thereby sign their own death warrants even more surely than they do by merely reproducing the virus, because although the toxin they manufacture might not damage the individual cell, it will kill the whole caterpillar.

Farmers may not be spraying the engineered virus onto their crops for at least a decade. The research will be followed by development of the product which must then pass stringent safety tests regulating the release of genetically modified organisms before commercial production can begin. We expect that a successful viral insecticide will kill caterpillars within 48 hours. It should be no more expensive than current chemical ones, and it will be vital that farmers use it circumspectly to avoid the risk of resistance developing. The over-use of many chemical insecticides in various parts of the world has led to a faster than necessary development of resistance.

If all this work comes to fruition as planned, the persistent, undiscriminating, and noxious chemicals of the past will slowly be replaced, and more efficient and chemically-free crops will be a dependable and renewable resource.

SECTION 3

Mining their own business

3.1

Environmental strategies, technologies and management in industry

IAN HORE-LACY, MIDHURST SERVICES PTY LTD

Ian Hore-Lacy is a biologist and industry historian who worked for twenty years in the mining industry. He joined CRA in 1974 as one of the first head office environmental scientists in industry and worked on a variety of site-specific and broader environmental questions. He is author of several books including Mining and the Environment *(4 editions),* Nuclear Electricity *(3 editions) and* Living in a Food Web *(3 editions).*

Environmental pollution by mining and manufacturing is as old as these activities themselves. However, the means to diminish that pollution and the motivation to do so have steadily changed. The technology currently available for avoiding many kinds of industrial pollution was just starting to come into use twenty years ago, and much of it was almost inconceivable twenty years before that. In those forty years we have moved from a position where environmental pollution was a peripheral consideration in the design of new plants to the situation today where it is a significant, and major, consideration.

Conservation and land rehabilitation have a longer track record, at least in Australia. The Victorian Soil Conservation Board (later the Soil Conservation Authority) was founded in 1940, and the first extensive mine-related revegetation project commenced in the mid-1930s at Broken Hill. Broken Hill was revegetated precisely because people continued to live there. They did not like the desolation brought about by clearing and overgrazing in that arid zone nor the effects of the dust which were exacerbated by the direct human disturbance of mining. (In Broken Hill the mine wastes themselves were not tackled until the 1950s.) Where mines were worked out and abandoned there was no such motivation for restoring the area. Mines like Captain's Flat near Canberra were abandoned in 1962, and some remote Northern Territory mines as late as 1971, without any rehabilitation. Some of these required subsequent treatment as changes in community attitudes, along with political considerations, became prominent.

These exemplify practical and tangible initiatives. At the same time Australia's population became increasingly urbanised and notions of the bush became less practical and more romantic. In the late 1960s scientific ecology was gaining prominence in school curricula, and environmental concerns—more often abstract than practical—attracted an increasing amount of media attention over the next two decades.

Irresponsible and neglectful attitudes began to change under the combined influence of environmental laws and regulations, and a new ethical view on conservation (or at least some measure of companies' self-interest). Extensive mining operations, for mineral sands and bauxite for example, pioneered post-mining revegetation in the 1960s; these practices spread to open-cut coal mines, and then more generally, in the 1970s. By 1980 post-mining rehabilitation was standard practice, although there was still plenty of scope for improving the techniques. However, thanks to the 1960s rehabilitation, in many areas along the east coast of Australia it is now virtually impossible to distinguish mined areas from undisturbed habitat. Some mined areas have now been incorporated into national parks.

While mining uses very little land overall (about one-fiftieth of 1 per cent of the continent's land surface) its local effects may be severe and this is reason enough for remedial action. In addition, the value of minerals gained from 1 hectare of land disturbed for mining is usually enough to make proper rehabilitation affordable. The most recent balance sheets of some major mining companies include the following figures for site rehabilitation: BHP $588 million, CRA $88 million, WMC $471 million, Pasminco $110 million, NBH–Peko $51 million.

The broader question of how to determine land use for all kinds of development activities has been more contentious, and during the 1970s this was tackled substantially through the Environment Impact Statement (EIS) process. Theoretically this was dealt with according to each state government's procedures for land use planning, but from the early 1970s any development proposal was liable to environmental impact scrutiny, forcing greater and more formal consideration of the conservation implications.

One spin-off from all this is that the mining industry is now reputedly Australia's largest non-government employer of environmental professionals. They are engaged in research as well as operations. This in turn has some commercial implications in selling Australian environmental expertise abroad. Industry is also responsible for a significant amount of basic biological research on contract, in connection with proposals for new projects.

Australia has largely been spared the widespread industrial pollution of other parts of the world, but only because we have had few dense concentrations of industry. Certainly, until the 1950s industry threatened the quality of life of neighbouring areas, especially those downstream or downwind. The 1960s brought the beginning of change and by the 1970s the need to reduce pollutants in emissions and effluents was widely recognised.

Prior to this, some known health hazards such as lead in smelter emissions had been tackled and much reduced since the 1920s, with steady but less dramatic reductions continuing as technology improved and regulations tightened. Fugitive emissions, particularly dust and fumes, were part and parcel of many industrial operations. They were all too conspicuous, at least until the 1950s, and particularly a problem with heat processes. Occupational health considerations and clean air legislation made these increasingly unacceptable, and as plants were replaced, cleaner conditions became more widespread. For older plants at least, the level of ventilation was increased, drawing dust and fumes away for cleaning in scrubbers or filter bag-houses.

Process emissions in waste gas streams were dealt with at much the same time by using electrostatic precipitators, bag-houses or scrubbers. Waste materials such as sulphur dioxide were often concentrated enough (or newer process equipment could make them concentrated enough) to be recovered as product or by-product, such as sulphuric acid. Waste materials that could not be recovered economically and posed no threat to health were sometimes still emitted to atmosphere, but more often dispersed through tall stacks. Some sulphur dioxide from smelting fell into this category. Australia is fortunate in not having a sulphur dioxide problem from coal combustion, since its coal has a very low sulphur content. Furthermore, our atmospheric conditions are drier than those of Europe and North America, and do not cause acid rain.

Overall, from the 1950s to the 1980s there was a marked improvement in air quality with respect to all forms of industrial pollution. This came about through improved process technology, some completely new processes, better pollution control technology, and better operating procedures. It was strongly driven by occupational health and environmental air quality concerns.

Liquid effluents have similarly been reduced over the last 40 years or so. Here the impetus has been environmental rather than occupational health concerns, driven partly by a rapidly evolving regulatory environment. Gross water pollution became increasingly intolerable, and a wide range of technologies, from the simple to the sophisticated, were employed to clean up process effluents (including cooling water) and surface runoff from industrial plants. Once again, both effluent quality and receiving waters were monitored to keep the pressure on for change, which was widely achieved. The means employed correspond to those outlined for air quality improvement.

The cost of all this was sometimes considerable, but became increasingly difficult to identify separately from other cost factors, such as the energy efficiency of processes and improved operating efficiency. For some industries, the need for pollution control has influenced their choice of new technology, though usually other factors have also been involved. Add-on or 'end of pipe' pollution control equipment and procedures are generally much less satisfactory than an improved process, which is intrinsically more efficient. But all economic factors must be considered, along with anticipated regulatory or public acceptability aspects, and rational decisions must be made. While an enterprise may well prefer to close down rather than cause gross pollution, none is likely to agree readily to becoming unprofitable in order to reduce pollution by a few percentage points and so achieve a particular arbitrary standard, the level of which is often contentious.

In the USA and Europe industrial emissions levels have sometimes been made explicitly flexible by trading corporate permits to discharge pollutants. Thus an enterprise which finds it relatively cheap to reduce emissions of a particular pollutant will sell its rights to discharge to another enterprise, which for the time being cannot reduce its emissions to the required level. From the state's point of view, the required reductions overall are achieved, and so is cleaner air, but the economic load involved is shared equitably in the marketplace. Less sophisticated, ad hoc variations to achieve the same objective occur in Australia, where consent may be obtained to pursue the most cost-effective pollution control, while intractable sources are allowed to continue discharges at levels which would not be acceptable indefinitely.

Unrealistic demands sometimes arise when new technology in trace element analysis is applied to effluent streams. For example, when Greenpeace sampled the effluent from the NuFarm chemical factory it found detectable levels of toxins which were acknowledged to be serious pollutants. The fact that the substances were in parts per trillion, and occurred naturally in similar concentrations, was not brought to the attention of television viewers. (One part per trillion corresponds to the thickness of a credit card in relation to the distance between Earth and the moon.) By the time the matter had been investigated, and the company exonerated of any wrongdoing or environmental imposition, it had lost some $7 million.

In recent years the residual pollution of land areas from industrial occupation (often including some effluent) has become a major issue. It not only affects the real estate value of such land but may mean that a previous occupier or owner is liable for clean-up costs, which may be very high and amount to many times the real estate value of the land concerned. Heavy metal pollution (e.g. from lead, zinc, cadmium, mercury) is difficult to remove in such situations. If it is deemed hazardous, the only option apart from removing a layer of contaminated soil is to immobilise it chemically. But organic wastes, especially oils and aromatic hydrocarbons, can increasingly be degraded by bacteria; in recent years a bioremediation industry has begun contracting for this. The significance of it is not simply that of a new (or newly harnessed) industrial process, but it is a reminder to the public that natural processes can improve environmental conditions, given time. Commercial bioremediation works by providing optimal conditions to speed up the natural processes.

While the need to reduce or avoid pollution is widely accepted by Australian industry, there is debate about particular standards. Also, there is sometimes uncertainty about the best means of achieving desired levels of control, and behind the environmental issue is the constant need to maintain and improve competitiveness in a purely economic sense. This means that as standards tighten, it becomes more important that they are well-founded. Unrealistic or unnecessarily tight standards do not serve the interests of environmental quality in any broad sense because they push industrial processes offshore from our well-regulated environment to a Third World or similar situation where pollution is less regulated. Nor are unduly tight standards likely to be rational according to other criteria. Indeed, we must 'think globally and act locally' by allowing reasonable industrial development on our shores, and by being consistent in our environmental expectations in all spheres of human activity—urban settlement, communications, agriculture for instance—rather than being preoccupied with industrial development and mining.

The 'precautionary principle' must be applied to policy-making, land use planning and risk management, where it means making conservative assumptions about the variables that have to be taken into account when assessing projects or conducting operations.

When I joined the mining industry in 1974, part of my role was to visit sites and then make some, albeit guarded, comment to management on the environmental functions and practices there, especially on emerging problem areas. Almost twenty years on, that kind of role has been upgraded in expertise and authority as the environmental audit. This is an integral part of management systems for many companies. Specialist teams, either from the company or external, visit a site and ask questions about the various Acts and regulations under which it operates, then look at evidence of compliance, management, and operating conditions. The specialists then produce a report which may be limited to identifying and ranking issues, or may extend to making recommendations. In companies with highly devolved management authority, it is essential that site management 'owns' this audit report. In any case, the action plan arising from the audit findings can be used to measure progress with various issues. Finally, in companies such as CRA a Board Environment Audit Committee has access to the reports and advice, and a brief from which to press searching questions of senior management. This is the only function elevated to the same status as financial audits in CRA, so it is far from being a merely peripheral concern.

Over the last two or three decades there has been a steady improvement in the environmental performance of industry due to improved technology, changed attitudes and tighter regulation. The improvements have, in general, more than compensated for increases in industrial activity. Environmental rehabilitation is part of company balance sheets, indicating that environmental conservation has moved from a peripheral consideration to a central concern in industry, and environmental audits are widely established as a part of corporate self-regulation.

The remaining challenges are technical, human and scientific. However, a considerable amount of industry's research efforts are deployed to grapple with those challenges, and in the next decade further significant progress is likely.

3.2

Our natural resources—economics, ecosystems and exploration

The Australian Mining Industry Council

The Australian continent is well-endowed with mineral and energy resources which provide a national economic advantage. There is also great potential for further resources and for extending known deposits as our geological, metallurgical, engineering and other management skills increase. However, the basic geological principle is that the Earth's resources are where you find them, not necessarily where you want them to be. Lauchlan McIntosh describes the background and dilemmas facing the mining industry.

The mining industry is a collection of organisations committed to finding, developing, processing, transporting and marketing minerals and energy products. In the last few decades the industry has developed many techniques to ameliorate and repair its effect on the environment. It has established a significant network of practitioners and environmental managers in areas such as ecosystem research, land management, rehabilitation and pollution reduction. Besides these practical networks, research has been sponsored at universities, government agencies and other appropriate bodies. Initiatives include the Australian Centre for Mine Site Rehabilitation Research and the formation of the Australian Minerals and Energy Environment Foundation.

There has been an increasing recognition of the benefits of strategies for sustainable development and these, in turn, have led to more environmental research, improved management practices, economic development and improved technology transfer between various industry, research, government and community groups. A commonly quoted comment on resources is that of the US Geological

Survey and Bureau of Mines: 'Resources must be continually reassessed in the light of new geological knowledge, of progress in science and technology, and of shifts in economic and political conditions'.

Exploration and new discoveries often extend our understanding of the genesis of ore. They can lead to the revision or replacement of geological theory, and even to further discoveries which would not have been considered remotely likely a few years previously. Mining companies use contemporary exploration methods that enable mineral exploration to proceed to an advanced stage with little or no surface disturbance. Over the past three decades there have been improvements in the efficiency and effectiveness of analytical techniques, starting with slow wet-chemical methods, moving to atomic absorption spectroscopy, and on to the most recent developments in 'inductive coupled optical emission and mass spectra' techniques which rapidly measure in one step the concentrations of many tens of different elements in a single sample. Concurrently, geophysical exploration techniques have also greatly improved. A Perth-based partnership between industry and CSIRO is applying the latest remote-sensing technology in Australia and around the world using advanced image processing and enhancement techniques developed by CSIRO. The joint venture, known as Geospectral Image Services, has undertaken eighteen projects since it was established in 1989. These have ranged from locating potential iron ore mineralisation in the Hamersley Iron Ranges to preparing base map images of Sarawak. Most of these projects have dealt with mineral exploration problems. In one case, night thermal imagery from AVHRR weather satellites was used to locate ancient coastlines in inland Australia as a guide to possible gold deposits. In the Carnarvon Basin of Western Australia another project generated base-metal target images from Landsat data for follow-up investigation by geologists on the ground.

An advanced airborne scanner which can identify ground-based objects only 5 metres apart has been built by CSIRO and BHP to monitor the environment and explore for valuable minerals. The scanner, called an imaging spectrophotometer, is a remote-sensing instrument 20 to 40 times more powerful than the Landsat satellite measuring instruments that are used at present. The spectrophotometer has several applications for mineral exploration, environmental monitoring, monitoring of soil erosion, salinity, deforestation and overgrazing.

Another airborne exploration technique is the laser prospector developed by CSIRO. The eagle-eyed device used in low-flying aircraft helps mineral companies spot faint traces of buried mineral deposits. As the aircraft flies at 300 km/h, the scanning device paints 2-metre patches of ground with 100 different wavelengths of infrared radiation, generated by a carbon dioxide laser. The laser beam stimulates minerals present in the terrain to emit infrared radiation. Each mineral has a characteristic infrared signature, which can be sifted out of the spectrum of radiation returning from the ground to sensors aboard the aircraft.

Exploration is the basic research so necessary to find new deposits. The exploration process uses every possible modern technique to assess all the information the Earth's surface provides about its underground resources. Geochemistry, geophysics and geobotany are essential disciplines in the process. Satellite technology, highly sophisticated computer databases, and data manipulation allow comparisons of rock assemblages in ways previously unconsidered.

The most significant problem for the industry is that access to land for exploration is fraught with difficulties, despite the contribution of current exploration technology. At the time of writing, about 6 per cent of Australia is set aside as national parks or conservation reserves. The borders of most of these reserved areas are imposed by the boundaries of leasehold or private property, public roads, rail easements or reserves established for economic purposes. In isolated areas where these constraints do not apply, the boundaries are usually polygons comprising straight lines drawn on a map to coincide with cadastral markers or with latitudes or longitudes. Consequently, the boundaries of many parks and reserves in Australia consist of a series of straight lines. Few of them follow natural features or ecological boundaries since few were delineated according to optimal natural considerations such as ecosystems, habitat limits, water catchments or natural features such as ridges, rivers or geology. Many land types and ecosystems on the unreserved side of a park boundary are indistinguishable from those in the reserved area.

Examples of this approach to reserving land for conservation purposes include: Kakadu National Park in the Northern Territory; the Lawn Hill National Park in Queensland; the Rudall River National Park in Western Australia; and the Sturt National Park in New South Wales. It is interesting that many areas of natural beauty occur where mineral assemblages are also likely to be found. Many national parks in Australia have been sites of past and recent major mining activity. Some examples of these are the gold mines of Kiandra, now in the Kosciusko National Park; the coal mines of the Blue Mountains; the mineral sands operations along the coasts of New South Wales and Queensland; and Coronation Hill in the Kakadu National Park. With today's technology and capabilities, many rehabilitated mines have been returned to national park status and commended for their environmental standards—for example, the Myall Lakes and Bunjalung National Parks in New South Wales. Other mining companies such as Alcoa in the south-west of Western Australia have won international awards for their rehabilitation performance.

A survey of former titanium minerals industry sites along 1700 kilometres of the east coast of Australia has demonstrated how sensitive areas of coastal land can be successfully mined. After mining, the restoration and rehabilitation of habitats can lead to subsequent inclusion in conservation areas. Some 30 of the mined sites are now included in eight national parks and seven nature reserves.

Most of the land in this country has been allocated according to the economic demands of agriculture, water catchment, forestry and urban areas. Most Australian parks and reserves were established in the last 25 years and came rather late in the land allocation process. Although boundaries often cover thousands of square kilometres, some of them are arbitrary and do not enclose complete ecosystems. Many either straddle or run close to prospective or established mineralised regions. Few, if any, national parks or conservation areas in Australia were declared after having been assessed for mineral potential. The mining industry maintains that exploration and resource assessments, even in existing parks and reserves, must continue if the community is to decide how best to integrate the economic and environmental benefits desirable from environmentally sensitive areas.

The mining industry believes that governments must be convinced that greater flexibility is necessary for the placement of parks and reserves so that industry management can promote ecological integrity and scientifically delineated parks systems while still providing the economic benefits of mining. Improving the scientific basis for parks and reserves will require studies to identify the ecological characteristics and the status of land surrounding Australia's present parks and reserves. This would be a first step towards providing more appropriate land to replace areas which might be lost.

There is little value in creating new protected areas unless benchmarks are established for biodiversity status and unless measurements of improvements or reduction in this status are made regularly. Excision, with the excised area replace by land of similar or better ecological value from elsewhere around the reserve boundary, would allow economic returns without carrying out operations in the park or reserve. After the area has been mined and rehabilitated it may be feasible to reincorporate it into the park or reserve.

Many new techniques and information obtained from demonstration projects are being applied to other parts of the country. The restoration of the Healesville Sanctuary is one such project. The mining industry accepts that mineral activity may, after assessment, be severely restricted in or prohibited from certain parts of Australia because of their scientific, ecological, historical or archaeological significance or outstanding scenic features. However, to accept this the industry needs to feel confident that it is not being frozen out of the research of the sub-surface by inappropriate or simply arbitrary boundaries. The mining industry in all its forms has demonstrated in many ways its commitment to integrating environmental research with its activities. This research has benefited other industries and researchers in Australia.

Geological resources are the hidden treasures of the future. Attempts to restrict access to them are likely to reduce our ability to understand not only the sub-surface but also the surface of Australia. From a scientific point of view the protection of surface features will reduce knowledge essential to land management in Australia. The integration of environmental research is an essential part of the mining industry's future planning. As David Brown of the USA Bureau of Mines notes:

> To be wise stewards of the land, we need to know its various resource values, whether surface or sub-surface, organic or inorganic. Good stewardship requires the benefit of inventorying the land's hidden resource potential. We cannot make wise decisions about the use of resources that we don't know we have. In effect, exploration today gives future generations more information about the possible choices they have about how to best meet their material needs.

3.3

The integration of environmental research and technology in the mining industry

Case histories from CRA

The CRA Group is one of the world's largest mining, metals and resource companies. It employs about 17 000 people at its operations in Australia and overseas, and its diverse competitive resources include iron ore, aluminium, coal, diamonds, salt, zinc, lead, silver, copper and gold. Resource deposits not yet developed include zinc, coal, uranium, titanium, minerals, copper and gold. CRA's activities encompass Australia, Japan, New Zealand, Papua New Guinea, the USA, Europe and South-East Asia.

Besides exploration there are several issues concerning mining and the environment that require consideration, such as access to land, limited environmental impact at all stages of mining, and multiple land use. The process of post-mining rehabilitation involves more than rehabilitation after the fact. The mining process consists of integrated and controlled strategies designed to minimise environmental impacts from the exploration stages right through mine establishment, operations, and site clean-up. Mining companies aim to minimise and, where possible, avoid environmental impacts whenever the temporary use of land is involved. This includes avoiding sites of known physical, ecological or heritage significance, and minimising disturbances to the natural environment's surface soil, vegetation, wildlife, drainage and aesthetic features. The pollution of land, water and air by physical and chemical effluents is also taken into consideration and environmental management plans are implemented to ensure proper pollution control.

In practical terms this means pragmatic planning by mining companies in matters such as site access. Where possible, existing tracks are used and as few new ones as possible are created. New tracks, airstrips and seismic lines are designed to reduce the long-term impact that land clearance and the disturbance of soils makes on vegetation and wildlife habitats, and to avoid any unnecessary interference to natural drainage patterns.

There is no question that mining orebodies disturbs parts of the natural environment, but there is no reason why topographical scars need to remain on the landscape. It is now considered a routine part of business for mining companies such as CRA to restore the land to a condition similar to that which existed before mining. In some cases the rehabilitation of mined areas has restored them to a condition superior to their pre-mining state. For example, the rehabilitation of open-cut coal mines in the Hunter Valley of New South Wales has allowed more cattle grazing than was previously possible.

HAMERSLEY—A NEW AGE IN ENVIRONMENTAL MANAGEMENT

Each minesite has its own unique requirements and has to be assessed individually. As mining becomes more varied so do the challenges of rehabilitation. Hamersley Iron is a wholly owned subsidiary of CRA Ltd, Australia's biggest iron ore producer and the world's second largest exporter. It has had 27 years of experience in the Pilbara region of Western Australia, operates four mines there—at Mount Tom Price, Paraburdoo, Channar and Brockman—and is extending its operations into second-generation orebodies such as Marandoo. The Marandoo iron ore deposit is about 35 kilometres north-east of the town Tom Price in Western Australia. Following two years of extensive environmental research and exploratory studies, a proposal by Hamersley Iron to establish the Marandoo Iron Ore Mine and the Central Pilbara Railway was submitted to the EPA in 1991.

The Pilbara iron ore industry was based originally on large, high-grade orebodies supported by a local infrastructure. The development of these isolated, single orebodies was hardly discernible in the region. However, the rich unified orebodies are becoming depleted and since it is unlikely they will be matched by similar new developments it has become necessary to use smaller orebodies to maintain production and extend the life of the existing deposits.

The principles of managing the land are constantly evolving because what was considered good environmental practice only a few years ago is now being challenged. New approaches have led to new benefits in landcare, mining procedure and, ultimately, financial returns. The experience we have gained has enabled us to modify operations at our newest mine and to introduce a new age in environmental practice in the Pilbara mining industry.

This new knowledge was gained mostly from our environmental experience at the Channar location, where new studies led to significant developments in our standards and guidelines for rehabilitation at Marandoo. As these new principles are put into practice at Marandoo, Hamersley hopes to achieve excellence in rehabilitation and environmental management. It has, for example, developed expertise in managing arid lands. The Pilbara is a harsh, arid and isolated region that is slow to reflect the benefits of good environmental practice, and it challenges our present knowledge of the environment. While considerable data on various sites has been collected over many years, it has been relevant only for specific areas and not suitable for making comparisons throughout the entire Pilbara region. For example, computer modelling has generated potential final pit profiles to facilitate the design of overburden dumps.

The great challenge in successfully rehabilitating an ecosystem disturbed by mining lies in achieving a sustainable condition as close as possible to its original state. This can be achieved only when the components and processes of the original ecosystem are fully understood. This involves many scientific disciplines, from the technical to the ecological. It also demands flexibility so that changing circumstances can be met with technical innovation and the further development of adaptive environmental management. This flexible approach has helped to minimise environmental impacts and increase operational efficiency.

From the outset, Hamersley Iron has been committed to a responsible approach to the environmental management of the Marandoo Project. We voluntarily undertook an Environmental Review and Management Program (ERMP), the highest level of environmental assessment and also the most public. The environmental impact assessment of the Marandoo project involved a multi-disciplinary team of over 60 scientists and 21 major consulting groups, at a cost of $6 million. In pursuing this course Hamersley Iron found that its commitment to quality opened up many new and valuable opportunities. It gained not only good environmental practice but also cost savings and other efficiencies that benefited science and encouraged the environmental practice of others.

Work began in 1991 and proceeded along two main fronts—environmental studies, and the development of standards and procedures for rehabilitation. These came together with the completion of the ERMP early in 1992 and continued through that year as the future environmental management and rehabilitation program was planned. It was during the preparation phase for the environmental studies that our commitment to quality was tested. The second generation of Pilbara iron ore deposits presented a different challenge because they are smaller and require more infrastructure. A new approach to data collection was called for in preparing management plans.

In the past, biological studies in Pilbara had generally been restricted to an in-depth study of a specific area. In many cases this study provided the only source of information on the native fauna and flora species in these remote areas. Little was known about contiguous lands or how the site fitted within the broader regional context. This was further complicated by the fact that previous botanical data had been intended primarily for compiling species lists and preparing vegetation maps. The variety of techniques used to collect the data made comparison between surveys difficult. A regional approach to data capture and analysis was developed for integrating surveys. This emphasis on a systematic approach in a regional context was a departure from our previous practice and made a significant contribution to the quality of the results.

Another test, and one of the key issues raised during the early stages of the ERMP, was to develop a set of standards and procedures for adaptive environmental management and rehabilitation that specifically dealt with the environmental conditions of central Pilbara.

Substantial effort was made during the ERMP process to review the rehabilitation principles and practices currently accepted in Western Australia. This review clearly indicated that some of the rehabilitation theories and techniques used with great success elsewhere in this state could not be applied to the environmental conditions of the Marandoo area. This view was verified following a review of the rehabilitation of the areas disturbed during the construction phase at Channar. Several factors crucial to the Pilbara were identified, including the progressive operation and rehabilitation of disturbed areas from the outset of the development, the careful management of surface materials (such as topsoil, litter and remnant vegetation), and the careful application of land-forming and water-harvesting techniques. It was concluded that the diverse and dynamic nature of these environmental conditions demanded site-specific rehabilitation. Previously, rehabilitation prescriptions were standard rather than modified to suit the conditions at each site. For Marandoo we developed specific standards and guidelines to suit the conditions there.

Though sufficient to assess the impact of the Marandoo Project, the environmental studies completed as part of the ERMP process highlighted the need for further research into selected aspects of the biological environment. Consequently, in 1992 an extensive research program examined selected aspects of the components and dynamics of the natural ecosystems at Marandoo. It focused on identifying and examining selected habitats for flora species of special interest. This included delineating their habitats and collecting data on their populations. A similar study was carried out for the Pebble-mound Mouse, *Pseudomys chapmani*.

Marandoo will be rehabilitated progressively during the construction, operational and mine closure phases. The current program of environmental research and the development of standards and guidelines for rehabilitation involves the first two phases, which will run concurrently. New knowledge gained in this first phase of research will improve our land management and rehabilitation skills.

WEIPA—REHABILITATION

CRA has almost 30 years of rehabilitation experience in the bauxite mining industry at Weipa in North Queensland. In 1966, when most Australians were only vaguely aware of environmental issues, Comalco (67 per cent owned by CRA) was beginning its regeneration program at the Weipa bauxite mine in far north Queensland. Comalco has spent considerable time and money, and conducted extensive research on land rehabilitation at its open-cut mine at Weipa, and the program there has become one of the most outstanding examples of its kind in Australia and the world.

Since the program began, 4740 hectares of mined land have been revegetated, with 80 per cent of it returned to native flora such as local species of eucalypt and wattle. Every year about three-quarters of a million plants are introduced to revegetate another 280 hectares of mined land. Aborigines from the nearby Napranum community collect seeds for the program. The seeds are dried and then sown directly onto local regeneration areas or else seedlings are raised for planting in other areas.

The plants used for regeneration must be able to grow in Weipa's infertile soils and must have a high resistance to flood, fire and termite attack. Planting involves laying fresh topsoil, preparing the seed bed, and aerial seeding—this has the advantage of spreading seed rapidly over large areas immediately after rain has fallen, when ground conditions prohibit the use of a tractor. Some plants do not establish themselves easily and only produce a few seeds. To reintroduce these plants successfully the seedlings are raised in either the company or commercial nurseries and planted out as tube stock. Over 60 000 seedlings are raised each year for rehabilitating areas mined for bauxite.

When Comalco first started its rehabilitation program in the 1960s there was little information on the subject. Since then, 41 forestry trials, 17 crop trials, 13 pasture trials and 42 native flora trials have been conducted at Weipa. The aim is to establish self-sustaining native vegetation that will support native fauna, and to develop land uses for commercial industries.

MARY KATHLEEN—INTEGRATED REHABILITATION

The Mary Kathleen site has been successfully rehabilitated by Minenco, CRA's project management and engineering company which provides technical environmental support and services to operations throughout the CRA Group. In a remote north-west Queensland area about 60 kilometres east of Mount Isa, the Mary Kathleen mine ran in two phases for a total of eleven years. Mining commenced in 1958 and concluded in 1982. Employees were accommodated in a self-contained town of some 1000 people 6 kilometres from the mine. At the close of operations in October 1982 the mine site and town contained 1.8 million cubic metres of toxic liquor covering an area of 60 hectares; about 400 000 cubic metres of ground contaminated with this toxic liquor; 7 million tonnes of radioactive tailings covering 28.5 hectares; and 24 million tonnes of low radioactivity waste rock from the mine. The mine infrastructure consisted of a powerstation, workshops, offices, the open-cut pit, earthmoving and mining equipment, and the town itself.

The rehabilitation of a closed-down uranium mine was a sensitive issue. Our objective was to transform the mine site into a safe condition consistent with future land use (grazing and tourist traffic) and requiring no maintenance, before relinquishing the leases on it. The rehabilitation program, with a budget of about $17.5 million, included removing all structures on the mine site and in the town which would deteriorate and become unsafe, unsightly or require maintenance. For public safety, radiation and physical hazards were removed by neutralising or evaporating the toxic liquor in the evaporation ponds and the contaminated water below the ponds, and by monitoring radiation and radon gas levels across the site. All radioactive material was buried to keep radiation and radon gas emissions at acceptable levels. This required monitoring and modelling ground and surface water movement and chemical quality and adopting contamination standards. Natural revegetation was encouraged. Surfaces needing revegetation such as the tailings dam, evaporation ponds, waste rock dumps and general site areas were assisted with slope stability checks, erosion control, topsoil placement and the sowing of seed.

A key feature of the rehabilitation was the short monitoring period between completion of work on the site and the Federal Government's acceptance of responsibility for the leases. When the project was completed there were no physical hazards remaining on the site and chemical hazards had been reduced to acceptable levels. The migration of chemical hazards through ground and surface water had also been reduced to acceptable levels and future trends had been assessed.

Our follow-up surveillance has confirmed the success of the rehabilitation—for example, the movement of iron, manganese, radium and uranium in the groundwater flow from the tailings dam is negligible. Sulphate, however, is more mobile and has been used to check the modelling predictions made when rehabilitation began. The available measurements give a fair correlation with the model and indicate that long-term predictions for water quality can be met. The environmental impact of the mining operation on the only feasible economic activity in this region (stock grazing based on bore water supplies) will be negligible. However, in the short to medium term, residual surface water quality immediately below the tailings dam will still be above desirable levels for stock water.

As a result of successful completion of the project the relinquishment of the lease was accepted in April 1987. The project is a benchmark for the site clean-up of other mine sites in Australia and overseas, particularly in the uranium industry.

3.4

The rehabilitation of mine sites—taking the initiative

A CASE HISTORY FROM RENISON GOLDFIELDS CONSOLIDATED

Renison Goldfields Consolidated (RGC) is another mining company that has accomplished some of the best rehabilitation projects internationally and has spent years developing a variety of environmental strategies and technologies. Renison Goldfields Consolidated Mineral Sands (RGCMS), an RGC Division, mines mineral sands at Capel, Eneabba and Geraldton on the west coast of Australia, and at Green Cove Springs in Florida, USA. It is a major producer of mineral sands, with the capacity to supply 25–40 per cent of the world market, and it plays a major role in Australia's economic development by exporting rutile, synthetic rutile, ilmenite, zircon and monazite. RGC, a major Australian mining house, mines gold, copper, coal, and tin, in addition to mineral sands, in Tasmania, Western Australia, New South Wales, Northern Territory, Indonesia, Papua New Guinea and the USA. RGC is a pioneer of the mineral sands industry in Australia—since 1934 it has operated extensively along the coast of NSW and Queensland.

Mineral sands orebodies were deposited by the sorting action of waves and wind which usually resulted in shallow, narrow, linear orebodies or lower-grade high dunes. The heavy minerals are recovered by either wet dredging or dry mining methods, and are separated from the light sands and clays (tailings) by wet gravity spirals. The heavy mineral concentrate is transported to a dry mill where it is separated by gravity, electrostatic and magnetic methods.

RGC has a healthy respect for the environment. Few people realise that it has had strong environmental policies for over twenty years, including an endeavour to reflect changing community expectations and exceed statutory requirements. Its commitment to improve its environmental technology is maintained by the Board of Directors, and infiltrates all levels of management through several mechanisms. Environmental matters are a day-to-day part of doing business and range from rehabilitation, waste management, staff training, research, community relations and industry affairs.

RGC holds regular environmental training seminars for environmental and operations personnel. The overall theme of these seminars is the integration of environmental management into planning and operations. They work hard at communicating environmental information throughout the company and to the community. Our environmental staff are active members of local landcare groups in all RGC mine districts.

The mission statement of RGCMS is short and to the point: '. . . to continue to supply the world with mineral sands whilst protecting our valuable environment for future generations, thereby achieving a balance between conservation and development.' RGCMS takes those words seriously and is initiating hard-working landcare projects that far exceed statutory requirements. An example of this is the Australian east coast operation, for which there were no clear government guidelines.

North Stradbroke Island—dune Rehabilitation

On North Stradbroke Island, a large sand mass about 40 kilometres offshore from Brisbane, 32 kilometres of the island's frontal dune system were mined and rehabilitated between 1970 and 1978. Before mining, the dune system was largely unstable and devoid of vegetation. Wind erosion carried large volumes of sand landwards, away from the beach system, progressively burying a freshwater swamp behind the dunes. During storms, the beach and dunes were further eroded by wave attacks. Rehabilitation required reconstructing the dunes into a moderate frontal slope of about 10–15°, a well-defined dune crest and a varied topography to allow for a range of micro-habitats. This was achieved by the strategic placement of tailings by using hydraulic stackers at the end of the mining plant. The next step was to stabilise the newly formed surface with local vegetation that could withstand the rigours of wind, sand and salt spray, in conjunction with brush-matting and sand trap fences. The company developed techniques to harvest, process and sow large quantities of sand spinifex grass (*Spinifex sericeus*), along with a purpose-built seed drill, and up to 16 000 kilograms of spinifex seed were harvested annually.

A series of wetlands evolved by natural progression in the depressions formed in the frontal dune system, helped along by planting a rhizomatous sedge as a primary coloniser. The thick, pure stands gave way to many other native species, allowing the habitat to develop further. The biomass and organic accumulation increased quickly and soon provided food and protection for a variety of fauna including birds, reptiles, mammals, amphibians and invertebrates. The technology

RGCMS developed for Stradbroke Island has been used overseas and has been rated by some international authorities as the best of its type.

JERUSALEM CREEK—RE-ESTABLISHING A NATIVE ECOSYSTEM

At Jerusalem Creek in New South Wales, an area of 13 square kilometres was mined and rehabilitated to re-establish a complex native ecosystem. The area supports diverse drainage and vegetation patterns comprising dry heaths, wet heaths, sedgelands, and swamp woodlands. It was necessary to re-establish ground levels accurately. In this type of area variations in contour are subtle, and changes in the ground level as little as 15 centimetres can profoundly affect the drainage and vegetation patterns. It was crucial that the tailings were carefully placed. It was also imperative that the surface water was removed from the site before mining so that the topsoil could be stripped in a moist but not sloppy condition. Draining of the site after mining is also important to keep the water levels at or below the ground surface while the vegetation gets a chance to re-establish itself. During the summer months wetland vegetation in the establishment phase can be scalded by inundation as shallow surface water heats up. Drains are then removed as the vegetation becomes mature enough to cope with the conditions. Experience has shown that wetland vegetation establishes itself and grows best during dry weather, when water levels fall below the ground surface and soil moisture is still adequate.

Even the topsoil had to be handled with care since a high proportion of wetland vegetation recovers from vegetative propagules in the soil. Not only does the moisture content have to be considered but it must also be carefully handled by the bulldozer operators to minimise disturbance to the propagules. In this case the top 20–25 centimetres of soil was removed in slices to retain the soil and plants in large clumps, and stored to one side, while a further 10–15 centimetres was then stripped and stored separately. As soon as the mining plant returned the tailings to the mine path, the topsoil layers were returned to their respective positions.

Two rare species have recolonised the rehabilitated land. A small shrub, *Baeckea citriodora*, which is known to respond to disturbance, has regenerated profusely in the wet heaths, and the rare ground parrot *Pezoporus wallicus* has recolonised the wet heaths and sedgelands. This parrot has very specific habitat requirements which unfortunately have diminished elsewhere.

ENEABBA—ACHIEVING A SELF-SUSTAINING ECOSYSTEM

RGCMS's largest mine is on the dry heaths of the sand plains near Eneabba, 300 kilometres north of Perth in Western Australia and 30 kilometres inland. This region has long, hot, windy summers, with most of its 530 millimetres average rainfall falling in June, July and August. The soil tends to be sandy and infertile, with diverse dryland heath vegetation characterised by a high level of species richness and endemism, and a low order of dominance. There is no permanent surface water. A vegetation survey on a 20 square kilometre area covering the orebodies recorded

429 species, 15 per cent of which could not be identified to species level, 50 families and 162 genera. About 10 per cent of the species were classified as rare or poorly known.

At Eneabba RGCMS's strategy is to achieve a self-sustaining ecosystem that is far more extensive than the completion criteria set by the regulatory authorities. It is the most difficult site rehabilitated by the mineral sands industry so far. Mining commenced in 1975, and over 1000 hectares now under rehabilitation are increased by about 100 hectares a year. Basic rehabilitation techniques had to be adapted to the harsh conditions and complex vegetation. Three main seed sources are used, all of which contribute to the return of native species: topsoil; mulch containing bradysporous, 'canopy-stored', seed; and direct sown seed. Monitoring them has shown that the species regenerating from the mulch and direct seeding tend to dominate in the first year. Generally the bradysporous species regenerate in the first or second year, whereas the soil-stored species regenerate over a longer period.

Much time is spent on investigating the basic phenology and germination characteristics of specific plant families or species since little research on them has been conducted anywhere. The rare and endangered species of bearded heath (*Leucopogon obtectus*) was extremely difficult to propagate from seed or from tissue culture. Researchers at Kings Park were contracted to develop new techniques for propagating this species and this has assisted the propagation of other recalcitrant species. Fauna studies have also been conducted, including surveys on the return of invertebrates, vertebrates and avifauna to dry heath habitats and on the return of ecological processes. RGCMS also monitors the progress of the vegetation in various ways—from visual inspection to botanical sampling for a computerised databank—to assist rehabilitation.

RGC's open-air laboratory at Capel

At Capel in Western Australia, the RGC Wetlands Centre is being developed as an open-air laboratory. Its Management Committee seeks a considerable amount of advice on wetland conservation techniques and management from waterbird and wetlands specialists. For example, the Royal Australasian Ornithologists Union (RAOU) is the project manager, and scientists from three West Australian universities conduct research at the site.

Capel is an outstanding example of how industry, government, universities and conservation groups can work together to develop techniques for habitat rehabilitation. Wetland ecosystems are an integral component of the environment, but this valuable resource is diminishing; 75 per cent of all wetlands on the Swan Coastal Plain in Western Australia have been degraded or destroyed for other land uses. The Centre is a company initiative and is not a normal mining lease rehabilitation requirement. It is RGC's way of helping to reverse this trend by creating an open-air laboratory and wetlands centre on land that was previously a pine plantation.

During the early 1970s part of the State Forest pine plantation was mined for heavy mineral sands. The volume of sand mass lost in this high-grade area resulted in a series of lakes, while the surrounding land was returned to conform with the pre-mining contour. In the late 1970s and early 1980s these pits were used as reservoirs for mine water. In 1984 a study of the artificial lakes was commissioned to assess their potential for supporting waterbirds. The study indicated that the lakes were suitable for self-sustaining wetland ecosystems for the conservation of waterbirds. The company provided research into the wetland ecosystems as well as the development and management required. It also created facilities for public recreation and education, including the means of demonstrating wetlands rehabilitation technology.

Although birds tried to breed at the site they had little success because they did not have enough shelter from predators nor enough aquatic animals and plants for food. Baseline data was established for the lakes to determine the development and future research needed to create a self-sustaining ecosystem, and this led to an extensive program of habitat modification and research focusing on the food chain. The multidisciplinary strategy encompasses water and sediment chemistry, algae, fringe plants, aquatic invertebrates, reptiles and birds. This valuable information on individual species and their interactions with other facets of the ecosystem provides mining and other industries with new techniques for future rehabilitation projects.

Over 50 000 trees have been planted in the Capel wetlands. In the wet woodland an 8 hectare low-lying sandy area adjoining the lakes (but not connected) was excavated to just below the winter water level, surrounded by a moat to limit predators, and planted with melaleuca and other fringing plants. The aim is for this area to provide a breeding site for colonially nesting birds such as ibises, cormorants and egrets. It will act as an extension to a nearby lake on private farm land which is one of only two breeding colonies of the great white egret in Western Australia, but which is unprotected.

In the woodlands different communities are being developed to maximise the habitat's diversity. Melaleucas are being planted on the fringes of the lakes, and flooded gum seedlings are planted with grass and sedges to provide breeding areas for tree-nesting waterbirds. She-oaks shelter the lake, and tea-trees, wattles and banksias are being planted near the interconnecting creeks to attract stream zone birds, firetails and rails.

In the land surrounding the wetlands, the establishment of windbreaks in a mosaic of plant communities creates four structural types: forest, woodland, heath, and grassland to attract birds, mammals and reptiles. Several nesting boxes have been installed and a brush-tailed possum was one of the first inhabitants. Research into the reptile population found that the vegetation of the uplands is a critical link in enabling reptiles to colonise the wetlands from surrounding areas.

As part of the communications program, tours of the RGC Wetlands Centre are available on request, and up to 4000 people tour the Centre each year. Exhibits of rehabilitation techniques are displayed at local shows in all districts where we operate, while the RGC Wetlands Centre promotes waterbird conservation. Talks are given to local community groups, including Rotary and Rotaract to promote the technical knowledge gained and to convey it to the broader community.

RGC WETLANDS CENTRE

An important new initiative is the development of the Centre as an educational resource. The Centre is working with the Science Teachers Association of Western Australia to develop curriculum material relevant to the site. Professional development courses are now being conducted at the Centre to instruct teachers on the educational opportunities it offers, and teachers now use the Centre as an outdoor classroom for their students. In fact, some local schools have adopted the Centre as class projects and visit several times through the year for various activities.

3.5

Strategic support for minesite rehabilitation

Dr Clive Bell, Director of the Australian Centre for Minesite Rehabilitation Research

The Australian Centre for Minesite Rehabilitation Research (ACMRR) was established in 1993 as a joint venture between the CSIRO Minesite Rehabilitation Research Program, the University of Queensland Centre for Mined Land Rehabilitation and the Curtin University of Technology Mulga Research Centre in conjunction with the Australian minerals industry through the Australian Mineral Industries Research Association (AMIRA).

The Australian Centre for Minesite Rehabilitation Research is an initiative led and funded by industry to address the industry's strategic rehabilitation research needs. The combination of knowledge and resources will lead to a better understanding of the processes affecting the sustainability of reconstructed landforms and ecosystems to both the industry's and the nation's advantage.

The Centre's goals include providing sustainable environmental solutions to minesite rehabilitation which are acceptable to industry, government and the community. This will be accomplished by becoming recognised as a research authority that has the appropriate scientific and technical foundations to enable industry and government to set acceptable standards. The centre is also designed as a networking and communications focus for minesite rehabilitation in Australia.

The mining industry, which encompasses metallic minerals, coal, oil and natural gas, spends an average of about $800 million a year on environmental activities. About $130 million is spent on rehabilitation and, at a conservative estimate, $5 million on research. For example, an Australian Mining Industry Council (AMIC) survey of twenty major companies found that environmental research expenditure in 1990–91 reached $27 million; although most of this, at $13 million, was on non-invasive or minimal impact exploration.

The great majority of this expenditure on rehabilitation and on research has been on site-specific and short-term activities. In the last few years the industry has become aware of the necessity to take on longer-term (five years or more) strategic environmental and rehabilitation issues. The research, services and programs will be built around industry concerns including tailings disposal and remediation, final voids, waste rock and spoil dumps, revegetation systems, acid mine drainage and waste minimisation.

Another advantage of the ACMRR is that it offers the opportunity to enhance the training of undergraduates and postgraduates and increase the quality of information and technology to professionals in the industry.

SECTION 4

The corporate environment

4.1

Turning rhetoric into reality—promoting ecologically sustainable development in the corporate sector

DR RAYMOND C. NIAS, DIRECTOR OF CONSERVATION, WORLD WIDE FUND FOR NATURE

Protecting the environment has assumed prominence in local, national and international affairs. According to the World Wide Fund for Nature (WWF), business—one of the major generators of economic activity—will need to pay more attention to environmental issues if it is to prosper or even survive. In particular, business will need to consider ecologically sustainable development as a corporate objective in itself.

Defining sustainability is by no means an easy task. There are many definitions available, ranging from narrow economic interpretations to broader concepts involving social values and ethics. However, a common central theme of sustainability is 'development that meets the needs of the present without compromising the ability of future generations to meet their own needs' (WCED, 1987). In defining principles of ecological sustainability, WWF adopts a broad framework approach, rather than a narrow definition. Optimal resource use is only one aspect of sustainability; social equity and cultural issues are also fundamental. In essence, WWF seeks to demonstrate that nature conservation and economic well-being are not in conflict but are mutually interdependent.

Sustainable resource use and the concept of sustainable development have at their foundation the limits set by the Earth's biological processes—in the words of the WCED Report, 'the ability of the biosphere to absorb the effects of human activities'. The principle of ecological sustainability therefore transcends a purely economic approach and attempts to integrate economics within the boundaries set by ecological constraints. Sustainability also emphasises the qualitative aspect of development, as opposed to the more traditional emphasis on quantitative growth assessed by measures such as Gross National Product (GNP). Growth implies a simple increase, whereas real human development must also encompass improvement in both material and non-material well-being.

The incorporation of ecologically sustainable development as a fundamental goal of economic and social decision-making is probably the greatest challenge that faces humanity. The scale of human numbers, resource use and waste production is now great enough to affect adversely areas of the physical and biological environment, such as the protective atmospheric ozone layer, or the world's supply of fresh water, in a way that would have been unimaginable to previous generations. A large portion of the world's biological diversity—genes, species and ecosystems—is currently threatened.

One of the fundamental factors that has led to the over-exploitation and waste of resources is the inability to value and incorporate in the marketplace the costs of such waste. Since biological diversity underpins much of our agriculture, industry and medicine, it seems that every individual has a stake in its conservation. However, the true value of biological diversity has not been recognised and accounted for in economic systems. The costs of the degradation of the biological resource, the process of extinction and the destruction of natural and other productive ecosystems, have been externalised. The costs of not conserving biological diversity (such as the expensive rehabilitation programs needed to redress land degradation) are generally passed on to others and, in many cases, are passed on to the poor and to future generations.

There are many actions that individual companies could take to assist a smooth transition toward the sustainable use of natural systems and species. Broad guidelines appear in publications such as *Caring for the Earth*, published by IUCN, WWF and UNEP, and the *Business Council of Australia's Environmental Charter*. Specific guidance is now an important part of the services that some major consultancy firms and business advisers offer to business. The message is that corporate managers should ensure that their company develops an environmental policy, and that it receives due recognition by the entire company.

The process of developing corporate environmental policies and monitoring their progress will usually benefit from including government environmental agencies and community interests from early in the formulation stages, through to monitoring and implementation. Specific targets should be developed as part of the corporate planning process. Many companies now routinely establish policies for matters such as the use of natural resources, the siting of plant and waste facilities, the management of property with minimal disturbance to native species, and even deliberately encouraging the creation of semi-natural ecosystems. It is vital that such policies become universal and are continually monitored; improvements can be made as new knowledge and technology become available.

All companies should be investing in research and development to improve their environmental performance, and setting targets for such investment and

Right At Weipa, Aborigines from the nearby Napranum community help collect native seeds for Comalco's regeneration program. *Below* At Weipa, neem trees have been planted on mine areas to assess their suitability in the regeneration program.

Above Used paperboard collected and baled for recycling into new paperboard. *Below* Latest technology equipment used to produce recycled paperboard.

Above Ferrous metal collected and sorted for recycling. *Right* Electrical cable baled and ready to be stripped for recycling for plastic and metal components. *Below* Stainless steel scrap.

Left The LED head and amorphous silicon drum: new generation technology overtakes the decade-old laser cartridge. *Below* Out with the old, in with the new technology: Kyocera's Ecosys is cheaper, more flexible and environmentally conscious.

Above A Ford Falcon fitted with an LPG fuel system. *Below* A Ford delivery semi-trailer testing the use of LPG in heavy transit vehicles.

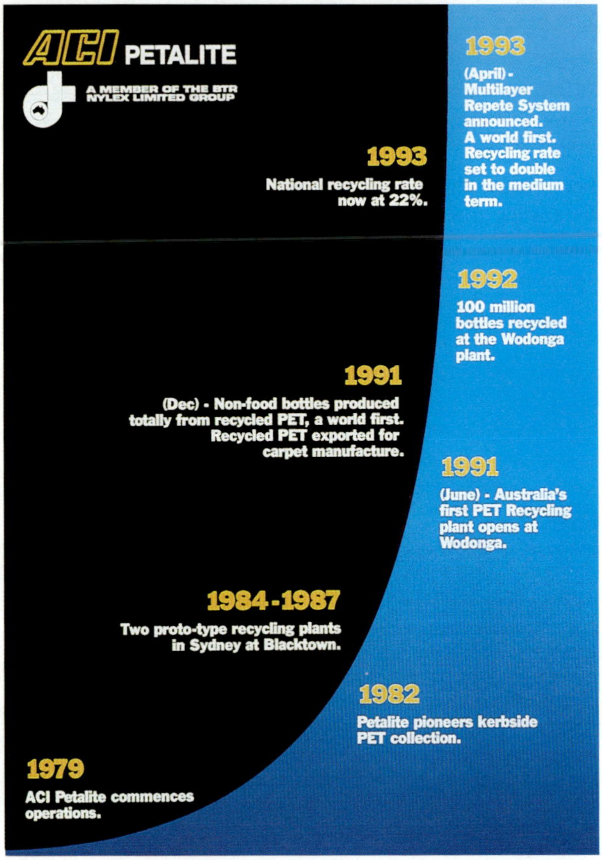

Top left Glass and plastic are collected for recycling. *Top right* Granulation: the first stage of the PET recycling process. *Above* ACI Petalite timeline: the historical development of PET recycling.

A clarifier, which separates the 'bugs' from the purified wastewater. The 'bugs' sink to the bottom and are pumped back to the oxidation channels to continue their important task. The purified water is then disinfected to kill any residual organisms, bacteria and viruses.

Biologists from the Department of Conservation and Natural Resources and Compol's environmental staff inspect a 500 g silver perch that has prospered in Lake Clements.

Valve monitoring: a plant operator checks one of the 1000 valves for leakage, to minimise hydrocarbon emission to the atmosphere.

Above A lagoon with land filtration in background. *Left* Werribee Farm—Lake Borrie lagoon: a 'Wetland of International Importance'.

performance, and these targets should be put on public record. If a company can be shown to have made minimal impact on the natural environment, it is possible to make a net gain for the environment by investing funds in conservation actions carried out by other agencies.

Specific strategies and plans to improve environmental performance can be based on an overall environmental audit of current practices—in the same way that financial audits can be used to highlight areas of deficiency or strength. The goal should be to develop Environmental Management Plans (EMP) that incorporate specific environmental goals and targets with timetables that can be subjected to annual audits and monitoring. Preferably, the results should be made available for public scrutiny. Given the importance and costs of such plans, governments should respond with appropriate incentives such as phased tax deductibility measures—for example, moving from 150 per cent deductions down to 100 per cent over a matter of several years.

The EMP should set out goals for matters such as using the best available technologies for waste reduction and pollution prevention, adopting reusable and recycled products, and reducing the waste of natural resources in the workplace. The environmental performance of contractors and material suppliers should be investigated as part of the plan and form the basis of an environmental purchasing policy. Such programs should be assisted by education and training programs for executives and workers, and considered part of the corporate culture. By acting as agents for positive change through such policies, companies can accelerate the move to sustainability and justifiably claim an important role in the development of policies and community leadership in the environment.

Clearly there is a lot that could be done, and clearly a lot needs to be done. However, there is every reason to expect that the private sector can and will adapt to and profit from such changes. History shows that commerce has survived momentous changes such as the abolition of slavery, the entry of large numbers of women into the workforce, the introduction of health and safety regulations, and much more. Change is inevitable: it will come either through community desire, or it will be forced upon us by the ecological limits of Earth's natural systems. Responsible industries will see the writing on the wall and decide to take advantage of new opportunities.

An important part of the change required for a transition to sustainability is open dialogue between business and the broad movement of social change (in this case specifically environmental organisations). Liaison between the non-government environmental movement and business is still in its infancy in Australia and much of the interaction on both sides can be considered reactive and confrontational. There is little evidence of progress in the development of consensus solutions. Progress may continue to be slow so long as non-government organisations remain critically dependent on reactive fund-raising techniques, with little opportunity for thorough analysis or policy development. There is no Australian equivalent to the large environmental 'think tanks' of North America and Europe. Research and policy development is notoriously difficult to fund, particularly as such activities do not normally represent the public relations return required by most corporate sponsorship.

There was an excellent opportunity to develop a strategic policy body during the operation of the Federal Government-sponsored ESD unit formed by the Australian Conservation Foundation and WWF Australia. This novel and creative partnership made a very important contribution to the ESD process. Unfortunately, the failure to attract funding for policy development work from outside government sources undermined its potential. The value of such an entity to the environment movement remains substantial, both to the movement itself as a provider of high-quality information and analysis, and to the corporate and government sectors keen to acquire alternative views on development issues. Such bodies could be charged with the development of new policies, evaluation and review of existing policies, and with providing consultancy services for government and the corporate sector. At present WWF is continuing to develop such relationships in the anticipation that the corporate sector will realise that long-term funding for policy development is a logical extension to the current emphasis on specific projects.

Funding remains a critical issue for non-government agencies attempting to deal positively with the corporate sector. Similarly, the ability of companies to market products or services that have an environmental premium will also affect the speed of positive environmental change. WWF is one of many organisations (including growing numbers of cash-strapped government agencies) that are attempting to raise funds from the private sector. In the case of WWF about 25 per cent of annual income is raised from private companies, primarily through project sponsorship or licensing arrangements using the widely recognised panda logo. In the manufacturing and services sector in particular, there are several promising developments as corporations seek to gain commercial advantage from 'green marketing'.

The extent to which 'green marketing' represents just another marketing opportunity, rather than a fundamental commitment to change, is still a matter for debate. Regardless of motive, however, an impressive array of more environmentally benign consumer products has appeared in the past five years. Such products, and others such as ethical investment funds or 'green bonds' should be supported—not as the solution to environmental problems but as a major boost to the development of 'green consumerism'.

Corporate funding, and techniques such as licensing arrangements, are often viewed with suspicion by the environmental movement. WWF conducts such relationships as part of an overall approach to corporate relations that promotes environmentally positive attitudes, services and products, not just a novel marketing opportunity. Ultimately there will be no alternative to a substantial reduction in the consumption patterns typical of developed nations, for both environmental and international equity reasons. Green consumerism can be seen as a step toward this realisation.

References

BCA (1992), *Principles of Environmental Management*, Business Council of Australia, Melbourne.

IUCN/UNEP/WWF (1991), *Caring for the Earth: A Strategy for Sustainable Living*, IUCN/UNEP/WWF, Switzerland.

World Commission on Environment and Development (1987), *Our Common Future*, Oxford University Press, Melbourne.

4.2

The virtue of selfishness in creating a sustainable environment

RICHARD PRATT, CHAIRMAN, PRATT INDUSTRIES

Pratt Industries has corrugated cardboard box making operations throughout Australia and a major operation in the USA. It also owns and operates paper recycling mills in Australia, the UK and the USA, generating sales that exceed $A1 billion. Pratt Industries says that it currently recycles the equivalent of more than 5 million trees a year. The privately-owned, Melbourne-based company now employs more than 3000 people worldwide and is Australia's largest producer of corrugated cardboard boxes made from 100 per cent recycled paper.

I believe profit is not only a positive goal, but one of the most powerful long-term incentives to ensure our resources and a regenerating ecosystem. Self-interest is a very strong motivator, and while this may be anathema to the young and naive, it is human desire that has brought the world to where it is today. That same desire and self-interest will determine the future.

The father of free enterprise economics, Adam Smith, recognised the power of self-interest some 200 years ago when he wrote: 'It is not from the benevolence of the butcher, the brewer, or the baker that we expect our dinner, but from their regard to their own interest . . . Never talk to them of our necessities, but of their advantage'. Nothing much has changed. Unfortunately the economic and political upheavals that are taking place around the world are more about self-interest than ideals. But this same profit motive has helped the world to prosper and to dramatically increase our standard of living. It has allowed the development of more (and more nutritious) foods, and the vaccines and technologies that allow more people to live better, healthier lives. As a nation, we have become richer, and as a result, we have become more environmentally aware. This has a lot to do with the fact that economic progress has given us the luxury of time to develop an environmental conscience.

Ideals are to be applauded, but the South American cattle farmers who are slashing and burning millions of hectares of rainforests for grazing land are mainly concerned with keeping their families fed, not the greenhouse effect. The higher living standards of the western world have fostered the development of a well-informed, highly articulate, and politically powerful environmental lobby. This in turn has led to a well-informed general public, which now demands that we deal with issues such as global warming, pollution, and the indiscriminate destruction of flora and fauna that may affect our children in years to come.

The environment is everyone's responsibility, regardless of political or philosophical leanings. However, it is interesting that many environmentalists cannot or will not see that there is a purpose common to those driven by profit and the environmentalists themselves who believe they are driven by altruism.

Unfortunately, turning back the clock will not work. The world already has too many people to support and will have to rely more and more on technology for food production, disease control, and clean water.

What is needed is environmental co-operation, or the development of environmental capitalism. Because like it or not, the interests of 'capitalists' and 'environmentalists' are inextricably linked, and industry and technology offer by far the best hope for improving the environment. Quite simply, the path that leads to ecologically sustainable development is lined with business opportunities, and the impetus will be the profit motive. This will keep our environmental future secure—provided of course, that consumers play their part. This is not just theory. A working example is that of the development of Pratt Industries, the company founded by my father and uncle. Like many immigrants my family tried its hand at farming, settling near Shepparton in Victoria to grow fruit. It did not take my father long to realise that there was a business opportunity making packaging to protect highly perishable products.

In 1948 they constructed a home-made corrugator and a boiler, and started making cardboard packaging. It started out as a humble, one-factory operation, trading under the name Visy Board. By employing good workers, by hard work, and by listening to our customers, we grew. By the end of the 1970s the company had moved into waste paper and cardboard recycling. This included building a paper recycling mill in Sydney to provide some of the packaging papers we needed to make our boxes. As the demand for the boxes grew, so did the company's expertise in waste paper recycling. More paper recycling mills were built to supply the increasing demand. Today there are five paper recycling mills operating in Australia.

The Australian paper mills make packaging paper for our box plants from 100 per cent recycled waste paper and cardboard. They consume more than 400 000 tonnes of waste paper and cardboard every year. To put it into perspective, the amount of waste paper and cardboard that the company recycles in Australia annually would be equal in volume to a 120-storey building. That is twenty storeys taller than the Empire State Building: more than twice the height of Melbourne's Rialto building. That paper would otherwise end up in landfill, adding to the already burgeoning amount of waste produced in this country.

The move to the recycled paper-making business over the last decade has made the company more than 90 per cent self-sufficient in supplying the paper needed to make boxes. In short, it is a billion-dollar business based on recycling. The company did not go into the recycled paper business for altruistic or environmental reasons, but solely for profit. It believed it could produce its own paper more cheaply than what it cost to buy it, especially when waste paper rather than trees could be used. The company responded to a demand: the motivation was profit. However, having gone into the recycling business through necessity, it has provided a perspective that has virtually limitless possibilities.

For example, in 1992 Pratt Industries opened the technically advanced $50 million de-inking plant and paper recycling mill at Coolaroo in Victoria, enabling us to de-ink up to 35 000 tonnes of office waste paper per year. The sophisticated use of computer control and the latest technology ensures a 100 per cent recycled product that performs and prints as well as packaging paper made from virgin pulp. The de-inked paper makes 100 per cent recycled white packaging papers used in the manufacture of white cardboard boxes—a product that the marketplace demanded. Another type of paper for which there is a demand is 100 per cent recycled de-inked newsprint, in preference to virgin paper. Accordingly, feasibility studies have commenced to make sure that enough of the product could be sold to make the project economically viable.

The company is broken up into three areas that cover the total recycling process: Visy Recycling collects waste paper and cardboard; Visy Paper converts the waste material into packaging paper; and Visy Board makes the 100 per cent recycled packaging paper into cardboard boxes and other packaging materials.

The process starts with collection. Visy Recycling's raw material is society's waste. It is the paper and cardboard discarded by offices and factories, shops and supermarkets, hotels and households. Regular collections in the city, town and country areas provide a steady flow of material to the recycling mills in Sydney and Melbourne. Pick-ups are made from commercial and industrial premises and from kerbsides by a fleet of company trucks, or by local council vehicles, contractors, transport operators, and community and charity groups.

The truckloads of material that would otherwise fill already stressed landfill sites start the process of being profitable products. The mountains of paper collected are bulky and awkward; many factories and retailers are supplied with baling machines to compact the waste, and 20–30 square metre containers for large volume sites. For smaller operations, a wool pack system is used. It is the customer's responsibility to separate the usable paper and cardboard from the other waste material because, for the recycling process to be technically and economically effective, pre-sorting has to be done at the source. This means that milk cartons, waxed cardboard, glass, plastic, polystyrene, wire, metal and other materials are removed before the paper is collected.

In paper-making it is important to distinguish between the different fibres that make up paper and cardboard. Paper products are made up from long and short fibres. Long fibres come from softwoods such as pine and fir trees; short ones from eucalyptus hardwoods. The long fibres give strength and flexibility; short fibres give quality. Waste cardboard is particularly important to Visy Recycling. For the end product a high proportion of long fibres is needed in the recycling mix—ideally 70 per cent long fibres and 30 per cent short fibres.

Visy Paper began operating in Sydney in 1980 with a small paper recycling mill, producing 40 tonnes a day of recycled packaging paper. From there it has grown to five paper recycling mills in Australia. Once the material arrives at the recycling plant, 11 tonnes of the waste material can be transformed, in an hour, into a 10 tonne roll of strong brown or white recycled paper.

Visy Board is the corrugated carton-making division. It is the company's flagship, and the business on which Pratt Industries was founded. Because of the virtually unlimited range of uses, the basic cardboard box is still an important part of Visy Board's business today. The increasingly sophisticated requirements of various customers have resulted in a wide variety of innovative package designs that not only perform their function but benefit the environment as well.

The environment is a major growth industry, and one of the fastest-growing industries in the world. When more companies begin to look at the environment as a potential source of profit, enormous opportunities will open up. Not just the switching of marketing strategies to promote existing products as 'environmentally friendly', but real environmental initiatives involving major investments.

4.3

The recovery of metal—the creation of profit

A CASE HISTORY FROM SIMSMETAL

Most metals are fully recyclable. In fact, the car you drive today may be part of the car you drove ten years ago. There are two basic types of scrap metal: ferrous, which is iron and steel; and non-ferrous, which is non-iron-based metal. By reclaiming valuable secondary metals, recycling helps to conserve scarce resources and the environment. In Australia, as in many other parts of the world, the recycling of ferrous and non-ferrous metals, such as lead and copper, is well established. The importance of recycling has received greater recognition over the past few years, and has drawn governments' and consumers' attention to its great potential for saving resources and energy, especially when this potential is compared with the costs of primary production. The contribution that recycling makes to the production of recyclable metals in Australia and around the world is also extensive.

The environmental savings made by using scrap iron and steel instead of virgin ore are well documented. Two of the major benefits are a 74 per cent saving in energy and a 90 per cent saving of virgin materials. Further environmental benefits are: less air pollution; less water use; less water pollution; and less mining waste. Similar results are achieved by recycling other scrap metal, and scrap metal products. Most metal products now contain a high proportion of scrap metal.

The scrap metal recycling process involves the worldwide collection, processing and trading of metal. European scrap requirements are supplied mainly by European scrap merchants and additional supplies come from North America. Supplies for the Asia–Pacific region come mainly from the USA and Australia, sometimes supplemented by European sources. The Asian ferrous scrap markets—principally Korea, Japan, Taiwan, Indonesia, Malaysia, Thailand and India—are net importers of 9–11 million tonnes of scrap metal per year, most of which comes from the USA, UK, Germany and Australia. As for most recycled commodities, the supply of free market scrap is driven by its price, the cost of its transport to collection points, and the general level of economic activity. For example, as economic activity increases, higher volumes of scrap metal emerge as manufacturing by-products and as a result of the replacement of plant and equipment. The construction industry—another significant source of scrap—is also affected by the general level of economic activity, because scrap sources from demolition increase as construction activity increases.

Ferrous scrap is a key raw material in the steel industry, and steel producers get it from two main sources. The first is called internal arisings; these are generated directly from inherent inefficiencies in the primary smelting and steel production processes. However, now that continuous casting and manufacturing has become more efficient the quantity of internal arisings as a source of ferrous raw material is declining. Internal arisings are generally retained and recycled by steel producers at their own facilities. The second source of ferrous metal is free market scrap which comes from consumer products, obsolete industrial equipment and the construction industry. Normally free market scrap is collected by recyclers who then sell it to steel producers.

Demand for ferrous scrap as feedstock for steel production varies according to industrial activity. In recent years the growth in electric arc furnace capacity, a production method particularly suited to scrap metal, has strengthened demand. While a basic oxygen furnace may typically use 25–35 per cent of scrap feedstock, over 90 per cent may be used by an electric arc furnace. The main sources of non-ferrous scrap are: domestic arisings such as plumbing fittings, copper piping, electrical wiring and aluminium cans; industrial arisings such as copper and aluminium manufacturing off-cuts and reject material; and shredder arisings from the shredding of automobile bodies and white goods, automotive parts including batteries and radiators and electric cables. Non-ferrous metal processing is largely confined to sorting, grading and processing scrap into uniform lots, consisting of like materials specified by customers, or standard industry grades. As with ferrous scrap, the volume of the international non-ferrous trade is determined by economic activity. However, since the intrinsic value of most non-ferrous metals is higher than that of ferrous metals, the volumes are not as sensitive to non-ferrous metal prices as ferrous volumes are to ferrous prices.

The principal use for non-ferrous scrap metals is in secondary metal processing for manufacturing. This includes ingoting and other semi-fabrication, and as raw material input for component manufacturing. Consequently the major demand for non-ferrous metals comes from countries with major manufacturing and industrial activities.

Simsmetal—Australia's oldest and most successful metal recycling company—was started by Albert A. Sims, a Sydney scrap dealer, in 1917. Simsmetal's operations, through its Sydney headquarters, are conducted on 46 sites in Australia, seven on the west coast of the USA, six in New Zealand, and one in the UK, with about 1100 employees worldwide and another 400 in Simsmetal joint ventures. Simsmetal is Australasia's largest collector and processor of ferrous scrap metal, handling over 50 per cent of the total amount of available scrap metal.

Simsmetal operates in three major business segments: the recycling and brokerage of ferrous scrap metal in Australasia, and in California and Washington State in the USA; non-ferrous scrap metal recycling and brokerage in Australasia and California; and the secondary processing of plastics and non-ferrous scrap metal in Australasia, directly for plastics recycling and aluminium, copper and brass ingoting, and through its joint ventures for lead smelting and copper and brass extruded product manufacturing.

The company currently handles about 180 000 tonnes per year of non-ferrous scrap worldwide. It uses its established network of sites for general ferrous and non-ferrous collections, but also uses special non-ferrous collection methods. Much of the non-ferrous material is sourced by individual metal type, requiring minimal sorting and grading prior to processing or sale. An example of this is the aluminium Cash-A-Can scheme in Australia, managed and operated by Simsmetal in association with Alcoa Ltd. The diversity of uses and types of non-ferrous metals and the relatively high value of non-ferrous metals mean that the material is collected from a wider range of sources than ferrous scrap is. Simsmetal obtains most of its scrap from spot purchases, period tenders, and contracts with manufacturers and utilities.

Simsmetal also operates three cable granulation lines in Australia, and one in the USA, which separate non-ferrous metals from their plastic sheathing. The plastic waste is usually sold to other plastic recyclers. It sells most of its non-ferrous arisings directly to the secondary processing industry, including its own operations. Although the proportions vary with economic conditions, about 45 per cent of Simsmetal's Australasian sales are domestic. In Australia, Simsmetal directs some aluminium and copper scrap to its own secondary processing operations. In New Zealand it trades most of its aluminium and copper-based scrap on the domestic market. The balance is used in Simsmetal's secondary processing operations in New Zealand, or exported.

Secondary processing involves melting, refining and ingoting certain non-ferrous metals. Although secondary processing is only a relatively small contributor to Simsmetal's operating profit, it gives the company strategically important flexibility in marketing its non-ferrous scrap, and significant value-added potential. Simsmetal's secondary processing operations are carried out in Australia and New Zealand, and the principal activities are the secondary melting of aluminium and copper alloys. During the past five years Simsmetal produced between 9000 and 15 000 tonnes per year of secondary aluminium and copper-based products, which amounts to about 20 per cent of Australasia's total production of these secondary metals. Simsmetal has effectively combined its market strength in non-ferrous scrap trading with significant penetration of the secondary processing of the market in Australia.

Simsmetal is also a major Australian producer and supplier of secondary aluminium alloy ingots and de-oxidant to the die-casting, foundry and steel producing industries. The company also merchandises a small volume of primary aluminium ingots, purchased from primary aluminium producers. The secondary copper ingoting facility at Milperra in New South Wales produces an extensive range of copper-base ingots for the Australian foundry trade for manufacturing a diverse range of products.

Simsmetal has interests in the secondary lead smelting industry through the Australian Refined Alloys joint venture (ARA) established in 1984 between Simsmetal and The Broken Hill Associated Smelters Pty Ltd, a wholly owned subsidiary of Pasminco Ltd. ARA, the largest and only significant producer of secondary lead in Australia, supplies ingots to the Australian domestic battery industry and to export markets.

In 1986 Simsmetal entered the plastic recycling business. In addition to the externally generated plastics that are recycled, Simsmetal can also realise the value of the polypropylene battery cases and the plastic cable sheathings generated from non-ferrous operations. The recycled plastic is sold primarily to the horticultural, agricultural, and building and construction industries.

4.4

Walking the 'thin green line'

A CASE HISTORY FROM KYOCERA

Now that social attitudes are changing, the demand for 'environmentally friendly' technology has become a challenge for many manufacturers. Kyocera is a manufacturer that has responded to this challenge. This case study looks at a major area of office technology and how one company has taken a revolutionary approach by providing a new generation of office equipment that is more economical and more 'environmentally friendly', and offers at least the same quality as the equipment it has replaced.

The enemy hidden in the office. Sitting quietly in the corner of most offices is an everyday machine that is no friend of the environment. This energy-inefficient machine is creating an increasing environmental waste problem, releasing a steady stream of ozone emissions and discouraging the growing trend of using recycled paper.

This unsuspected environmental enemy is the laser printer. It has been with us for a decade and until recently was applauded because it offered convenience and speed and made the printed page look attractive. The cornerstone of its design is the consumable cartridge—the engine of the printer. It was an ingenious idea that packaged all the unreliable facets of the machine—the drum, toner, developer and moving parts—into one removable, disposable plastic cartridge. However, its anti-social side-effects are beginning to cause concern: needless operating costs, environmental waste, energy inefficiencies, its inability to handle recycled paper freely, ozone emission and excessive paper use.

The economic and ecological disadvantages of the conventional laser cartridge are sixfold. The running costs of the machine are very high. Having to buy 'new' engines every 6000 pages or so means that it can cost three times more than the printer's purchase price just to keep it working. The environmental cost of dumping millions of these non-biodegradable cartridges in the environment is still being measured. More than 300 million cartridges will be produced by the end of 1994. The short-life organic drums in the conventional laser printer just are not tough enough to withstand the rougher texture of recycled paper. Paper accounts for 45 per cent of current landfill. Demand for recycled paper is forecast to quadruple by the year 2000.

Laser technology is bulky and inefficient in its use of energy and furthermore the technology emits ozone. While we need ozone in the upper levels of the stratosphere, we do not need it in the office: even the level of one part per million released by laser printers can cause eye and throat irritations.The great majority of printers cannot print on both sides of a sheet of paper. The few that do are unpopular because of loss of printing speed. Each year more than 115 billion sheets of A4 paper from countless trees are spewed out by office equipment.

What triggers the disposal of the cartridge is the expiration of toner, so virtually all the intelligent parts of the machine—usually still in good condition—are thrown out when the fuel runs out. It is like throwing away the car's engine every time it runs out of fuel. With laser printers, the most waste (and consequently the most cost) is incurred at the consumption or operating stage. The reasons why conventional lasers are wasteful, expensive to run and environmentally unsound all stem from the fact that most printers run on a cartridge-based engine. While the disposal of engines by incineration requires expensive scrubbing systems to prevent emissions of dioxins, particulates and metal oxides, their disposal to landfill is a waste of the world's scarce resources.

The present cartridge-recycling industry delays the dumping of cartridges to landfill for as long as possible, but even a recently opened plant in Australia can produce only 6000 recycled cartridges per month, while the marketplace is consuming and throwing away more than 1.2 million a year. Whatever the reason, independent analysts and environmental auditors claim that the level of reuse is depressingly low.

Today's recycled toner cartridge should have become an attractive alternative to the manufacturer's original but it has not. Many vendors discourage the use of third-party recycled cartridges in their printers, not least because profit from the sale of these consumables plays a substantial role in the whole pricing infrastructure because it subsidises the retail price of printer products.

Kyocera has taken a radically different approach. This company has simply elected not to make the mess in the first place. It has spent some years, and many millions of dollars, in fundamentally redesigning the page printer engine. Kyocera realised that the common denominator of the laser printer's high costs and its poor environmental performance was the high waste of the laser printing process and that very little could be achieved by fiddling with the laser printing design. The solution: to design an entirely new system—Ecosys.

The designers began by replacing the laser imaging technology with Light Emitting Diode (LED) imaging. This considerably reduced the size of the machine, its power consumption and the amount of moving parts in the process. Next, the designers applied Kyocera's knowledge of thin-film deposition technology to coat an aluminium printer drum with super-tough amorphous silicon. The drum, now a fixed part of the engine, could withstand 300 000 prints before it had to be replaced, which made Kyocera the only printer manufacturer to recommend the use of 100 per cent recycled paper. Coupled with a long-life developer, this revolutionary drum eliminated the cartridge from the printer's design as well as the environmental problems of its disposal. Finally, Kyocera took a modular approach to the printer's overall design, enabling users to build the system according to their own specifications. One of these models is a duplexing unit that enables double-sided printing.

With sales of more than $8 billion a year, Kyocera is the world's leading producer of ceramic packages for semi-conductor chips and is one of the top four suppliers of solar cells. It is a world leader in applied ceramics for metal replacement, and one of Japan's major telecommunications companies. It produced the world's first 18-page per minute desktop printer and the world's first laptop computers.

Our approach to business is based on our aim to learn constantly to become more environmentally responsible—to research and develop products that are gentler to our planet and place a lighter burden on the environment at every stage of their development, production, marketing, consumption and disposal. Kyocera practises what it preaches and must complete an internal environmental activity audit several times a year.

Kyocera sees responsible environmental management, like quality management, as a journey, not a destination. This journey we refer to as 'The Thin Green Line'—the narrow path between commercial reality and environmental idealism, where economics and the environment share common principles and can comfortably co-exist. Kyocera has scant regard for the marketplace clamour of shallow environmental claims. Any campaign to promote environmentally sustainable products that simply consists of a series of 'Buy green' bumper stickers aimed at the consumer will fail and the environment will be the loser.

Suppliers must build and explain the positive economic and functional benefits of environmental initiatives. They must commence the journey at the start of the product development cycle, not just when it is time to write the advertising campaign. Business consumers are not going to put their business' survival at risk by purchasing inferior, inefficient or expensive products just because they are environmentally friendly. The business must survive at all costs and survival in business means staying profitable. If staying in the 'black' means using environmentally unfriendly products because no other alternative is available, then who is to blame? Certainly not the business consumer. The responsibility for the acceptance and proliferation of environmentally friendly products lies with the designers and manufacturers, not with consumers. Harassing business consumers to buy 'green' to save the planet is as naive as asking people to 'Buy Australian' to save jobs.

DOES 'GREEN' SELL?

Kyocera considers that the ultimate marketplace test is the sale of Ecosys units. Since the printer made its debut in mid-1992 more than 5000 have been sold in Australia. According to Kyocera it took their share of the office laser printer market from less than 1 per cent to more than 6 per cent, and that means the elimination of more than half a million non-biodegradable cartridges which otherwise would have to be dumped in Australian landfills between now and 1997. Those purchasers of Ecosys will instead share an estimated saving of between $36–110 million in operating costs over the four-year life of those printers.

The independent research organisation, International Research Bureau (IRB) based in Sydney, investigated 'environmentally friendly' products to see if they offer users any real saving. The study was commissioned by Kyocera to examine the environmental impact of technology in the office by reviewing the manufacturers' initiatives for minimising 'office pollution'.

Author of the report, IRB Managing Director, Peter Hall says:

> We found that the efforts of the manufacturers to lessen the environmental impact of their products are beginning to flow through to actual products, centring on delivering 'green laser printers' with lower power consumption, higher duty cycles, duplexing for double-sided printing, better drum sensitivity to support all grades of recycled paper, and new engine designs to reduce waste by-products. Laser printers emerging with near-zero waste operate in much the same manner as yesterday's fountain pen, needing only toner refills. Kyocera has leapfrogged the market to emerge with a range of environmentally friendly laser printers that virtually eliminate consumable waste, through the introduction of long-life laser drum technology.

Green products such as the Kyocera Ecosys printer system, which support clean technology right back to the manufacturing process, have developed a 'green advantage' whereby the level of environmentally friendly development in the printer has led to an 'economic edge' in the marketplace. The report concluded that 'not only were Ecosys systems clearly the most environmentally responsible systems on the market, they were also more than 50 per cent less expensive to own than the average laser printer'.

The report has led Kyocera to believe that the greener the printer, the more functional it is and the cheaper it is to use and own. This should have major ramifications for office technology. The report clearly demonstrated that environmentally hostile products do generate significant cost penalties to the user. Many environmentally unsound products have generated huge profits for suppliers because environmental pollution was never treated as a production cost. This highlights the issue of the hidden costs that technology waste imposes on a community. The environment never charges when waste by-products are dumped on it—the cost of pollution is externalised from the supplier's balance sheet and put onto the user's shoulders. Indeed, we all become underwriters of these consequences.

There is no office technology that produces more environmental waste than a laser printer does. However, the disposal and cost of this waste falls on the user. By creating products that need the constant replacement of components, most laser printer suppliers have profited from environmental pollution. In testing the 10-page and 18-page per minute versions of the Ecosys system against all leading laser printer brands, the IRB proved conclusively that the more environmentally responsible a printer is, the cheaper it is to operate. It was the first report that was able to establish a relationship between the economic and environmental performances of office technology.

Kyocera has now turned its attention to the Small Office/Home User (SoHo) marketplace. In August 1994 we released a 4-page per minute, cartridge-free printer. True to its producer's philosophy, the new personal printer incorporates newer environmentally sensitive technology which Kyocera continues to develop. A drum durable enough to last 100 000 printed pages has been installed in the Ecosys 400 printer because the amorphous silicon drum would be unnecessary since it would take 20 years to wear out with average 4 ppm use. Kyocera engineers have succeeded in eliminating the need for a fan, with consequent energy savings. This also ensures the absence of the paper and toner dust that is normally blown into the user's environment by conventional laser printers.

Whether the printing speed is 4, 10 or 18 pages per minute, the Kyocera printer has been found to be the most economical, because it is the most environmentally sound; it is the most environmentally sound because it is the most efficient. This is the 'Thin Green Line'; this is environmental economics.

4.5

Driving the environment

A case history from Ford Motor Company of Australia Limited

Roger King, a Senior Engineer in Vehicle Safety and Regulations at Ford of Australia, is concerned with product-related environmental matters. In this case history he describes how the Australian automotive industry takes on the challenge of meeting emission control standards and still satisfying consumers.

Exhaust emissions

The main environmental issues affecting the Australian motor industry are exhaust emissions, fuel consumption, lead in petrol and chlorofluorocarbons (CFCs). The industry is regulated so far as vehicle safety and emissions are concerned by a series of Australian Design Rules (ADR)—53 in total—and ADR 37/00, the latest iteration of passenger car exhaust emission control, sets out to contribute to an improvement in the quality of air in our cities and to reduce 'smog days'.

In Australia, as in other countries with advanced industrial economies, car emission control targets unburned hydrocarbons (HC), carbon monoxide (CO), oxides of nitrogen (NOx) and emissions of volatile organic compounds (VOC) by evaporation. In 1986 Australia introduced catalytic converter technology for new cars. We estimate that something like 37 per cent of the cars in use in Australia now operate with this technology. The median life-span of a car is about 14 years, which means that the full fleet of cars will probably be recycled in 23–24 years. Thus in about 10–12 years most Australian cars will have catalysts fitted (and so will use unleaded fuel) and most will have 3-way catalytic action (for HC/CO/NOx control) with electronic engine management including closed loop control of air/fuel mixture strength.

Exhaust emission control demands are progressive. It is likely that by 2005, 30 per cent of cars will be certified to an even more stringent rule that will be introduced gradually from 1997. The Americans believe that a car with no emissions control at all will produce 54 grams of CO, 5.4 grams of HC and 2.2 grams of NOx per kilometre. At ADR 37/00 level, on a pollutant gas weighted average basis, our figures are just over 70 per cent better than a car with no emission control at all. When we move to the next iteration we will be more than 90 per cent better.

Ford, along with other manufacturers, has to provide passenger cars and light goods vehicles that are certified to ADR 37/00 but in doing so has to take certain things into consideration—longevity, for example. The emissions control system must still function so that the car emits no more than ADR 37/00 limits at 80 000 kilometres. This means you must be much better at 0 kilometres to ensure long-term compliance. Cognizance of production variability is important. There is a level below the maximum permitted that must be catered for to ensure that all production vehicles comply. To constrain the cost of your development fleet to a reasonable figure, the 0 kilometre target has to be even lower. So, at the start of its life, the average car is much, much 'cleaner' than required by the law.

Consumer research has found that the emissions control capability of vehicles has not been a particularly strong selling point. Emissions control is taken as given because, whether it is a big car or a small car, they both have to meet the same emissions levels although it is much easier for a small car to meet those levels. For example, people buy a Falcon (Ford's volume sales leader and Australia's most popular car) for its recognised attributes—a large car with plenty of space and towing performance. However, we are rapidly approaching the point where any further attempts to control emissions really start to follow a law of diminishing returns. The cost for the next level of control, depending on the car, will vary from $500 to $1500 extra on the price of the car. In the end the consumers will pay more for less improvement in emissions control.

Diesel emissions should also be mentioned. Until now developments have been slow in ADR legislation for diesel emissions control and all we have at present is a smoke limitation rule. Individual Australian states require that diesel vehicles shall not emit visible smoke for more than 10 seconds. ADR 30/00 (the smoke rule) does not have such a time limit, but has a smoke opacity requirement related to engine size and thus the throughput of gas.

People do not like to see smoky diesels; however, appearances can be deceptive. It is the much smaller particles, which car exhausts emit also, that affect the respiratory tracts of susceptible people. Because diesel engines run at 'oxygen rich' air/fuel ratios, and because current catalysts cannot chemically reduce NOx until the mixture has only just enough air to fully burn the fuel, NOx emissions from diesels cannot yet be controlled by catalytic action. Diesel engines are more efficient than petrol engines, and better combustion means that greater amounts of NOx are produced but less particulate matter is generated. Consequently the challenge for diesel engine designers is to reach an acceptable compromise between NOx production and particulate emissions. An additional ADR that will control emissions of HC/CO/NOx and particulate matter will be introduced in 1995.

ALTERNATIVE TRANSPORT FUEL

Alternative transport fuels, to replace petrol and diesel oil either completely or partially by blending, have been discussed in Australia for several years in the context of reduced exhaust emissions. These include Liquid Petroleum Gas (LPG), Compressed Natural Gas (CNG) and the alcohols ethanol and methanol. By far the most widespread in use is, of course, LPG. It is favoured by taxi drivers because of its low cost and good performance capabilities, but because consumption levels are higher than those of petrol and because it is mostly used in the city, LPG is considered no 'cleaner' than petrol. In relation to NOx emissions it can be much 'dirtier' if the engine is not tuned correctly for LPG. Ethanol and methanol are most suitably used in blends with petrol, and ethanol can be used with diesel oil. However, their capabilities for emissions improvement are questionable whilst other emissions (aldehydes) from these fuel substitutes are equally as toxic.

CNG looks like the best candidate for fuel substitution because it is 'cleaner' in terms of all the controlled pollutants. It is now being tested as a complete fuel substitute for diesel and in conjunction with diesel fuel and petrol. The problem is that it is a gaseous fuel and must be contained in pressurised tanks, occupying most of the boot space, in order to have sufficient for an equivalent journey using petrol. Ford has initiated a program to convert all of its 'kanban' (just-in-time trucks continually hauling engines, chassis parts and body panels between the Geelong manufacturing plant and Broadmeadows assembly plants) to dedicated CNG-fuelled engines.

FUEL CONSUMPTION

Alternative fuels, in the context of smog abatement, are also promoted as being less of a greenhouse gas concern as they produce less CO_2 per unit of energy. This is because the carbon-hydrogen molecule has lighter mass, in that it has less carbon. Again, this is what makes CNG an attractive substitute for petrol despite its lower energy density which reduces driving range. Methane (the main constituent of CNG), however, has greater global warming potential at a molecular level than CO_2.

Fuel consumption is presently a hot topic of discussion between industry and government because of the need to preserve Australian natural resources and the apparent need to decrease emissions of carbon dioxide and reduce the so-called 'enhanced greenhouse effect'. About 2.4 kilograms of carbon dioxide (CO_2) are released for every litre of petrol used. Ford is required, along with the other manufacturers, to report to government our new car corporate average fuel consumption (CAFC) for the prior year and our projections for the next consecutive three calendar years. We have done so since 1978 as the government responded to the oil crises of the 1970s.

During this time, Ford has improved the new car CAFC by about 27 per cent. On a national basis, it is about 25 per cent and the current Ford Falcon with the 6 cylinder 4.0 litre engine and 4-speed automatic transmission consumes about 20 per cent less petrol than the 1980 equivalent. This means that Falcon, the most popular car in Australia, now uses almost as little fuel as a 1980 4-cylinder 2.0 litre automatic Cortina. In late 1989 the automotive industry was asked to prepare a long-term strategy and in response the industry estimated that it could achieve about a 13 per cent reduction in fuel consumption between then and the year 2005, assuming that certain exhaust emission control enhancement actions took place.

People talk about fuel-efficient cars and immediately think small cars. It is true that in absolute terms a small car uses less fuel than a large car, but the Fairlane, which can tow a two-horse float or a boat, can use fuel more efficiently than Ford's small car, the Festiva, does. The important message about 'fuel efficient vehicles' is that it is efficient only if it meets the consumer's specific needs.

Ford engineers have helped make substantial cuts in national fuel consumption. Currently the new car National Average Fuel Consumption (NAFC) is something like 8.9 litres per 100 kilometres. The industry has assured government that by the end of this decade that figure will have dropped to 8.2 litres per 100 kilometres.

People are talking about carbon dioxide at the moment because more of it is released into the atmosphere than any other global warmer. Inevitably there is incomplete combustion (that is why there are toxic exhaust emissions). CNG may not produce as much smog, so it is 'cleaner', but it still contributes to global warming because of its methane content.

Even hydrogen, which is being promoted as the ultimate fuel 50–60 years down the track, may not be the solution because the most potent global warmer of all is water vapour, and when you burn hydrogen it produces water vapour. Thus water vapour from a car exhaust is emitted into the atmosphere. Water vapour from a car will linger in the atmosphere for only a few days, whereas carbon dioxide may remain for up to 100 years, and methane for up to 10 years. 80 years from now, though, when perhaps every car worldwide is using hydrogen fuel, global warming may still be a concern.

Currently natural gas costs about the same as LPG as there is no excise on either fuel, but the main problem with CNG is lack of infrastructure. Most urban refuelling outlets have LPG on their forecourts but similar infrastructure for CNG does not exist. Although CNG is a potential alternative it is not considered viable yet, except as a diesel substitute.

LEAD IN PETROL

Lead in petrol is still a major issue despite the fact that a leaded petrol car has not been manufactured in Australia since 1986. Essentially, lead is added to petrol as an octane-enhancing measure whereby designers can select a higher compression ratio for an engine to run more efficiently. However, this may be achieved in other ways—increasing the 'aromatic compound' proportion of the fuel specification, for example.

The oil companies operate according to set octane levels that vary in some countries. They say that the optimum octane for Australian unleaded petrol is 91, and so higher octane levels become less viable economically, whereas Europe has a 91-octane level, a 95-octane level, and a 98-octane level. If the same selection was available at acceptable prices in Australia, the problem could be lessened. Drivers could change to unleaded fuel overnight if their engines could function without a natural lubricant in the fuel to prevent valve damage.

In Australia about 20 per cent aromatics are used; some European countries use at least 40 per cent aromatics. This is a source of concern because there is the risk of exchanging one health hazard for another. High lead levels in the blood have been associated with impaired intellectual development, and benzine, an aromatic compound, is a carcinogen. The leaded petrol debate continues. The oil companies claim that they will drop to 96-octane for leaded petrol. That may sound insignificant but many engines will run the risk of detonation, which can seriously damage running gear such as pistons and rings, particularly if it happens for a marked length of time. The final impetus for the complete removal of lead will come from the oil companies as it will become less and less commercially viable for them to make leaded petrol.

AIR-CONDITIONING REFRIGERANTS

Chlorofluorocarbons (CFCs) are good for just about everything except the stratospheric ozone layer. The free chlorine molecules that leach out of the CFCs are said to damage the ozone layer. Again, we at Ford subscribed to the industry commitment to government and will have phased out the use of CFCs in new vehicle air-conditioning systems by the end of 1994, a year ahead of the Montreal Protocol requirements. Ford is also considering the development of kits that convert a vehicle's CFC airconditioning system so that R134a refrigerant can be used instead. R134a is a hydrofluorocarbon (HFC) that does not affect the ozone layer because it contains no chlorine. It is the same refrigerant that has been developed for airconditioning systems on new vehicle models introduced since the middle of 1993. Its characteristics differ from those of CFC and consumers converting CFC systems to R134a may notice a change in performance.

SECTION 5

Packaging problems and opportunities

5.1

The truth about recycling

LEIGH CATO

Everybody creates waste. Everybody has a theory for how to dispose of it. There are some who ask 'Why create so much in the first place?'. Needless to say each industry sees the problem from its own perspective.

Each year more than 14 million tonnes of solid domestic, commercial and industrial waste are disposed of in Australian tips or landfills. That is almost 1 tonne of rubbish per person, per year. An Industry Commission Report says that Australian cities are running out of dumping space. As the population grows, the problem becomes more urgent. Not only will individuals as well as commercial and industrial enterprises have to cut down the amount of garbage they produce, but better methods of disposal and recycling will have to be found.

To make recycling a widespread activity, it must be cost-effective, and the features and price of recycled products must appeal to consumers. The recovery of recyclable materials can significantly reduce the volume of waste going into tips and landfill sites, but first it must be made economically viable. Domestic recycling has been touted by many environmentalists. It is assumed that it is sound environmental practice and that savings are inevitable. Yet like everything else, domestic recycling is subject to economic rules and the benefits have to be weighed against the costs.

Recycling is commonly promoted as the way to save scarce resources for future generations. This sounds like a good idea, but it is a simplistic one, since recycling does not always save money, energy or other resources. For example, if we recycle something that is plentiful by using materials that are less plentiful, the process will inevitably be more expensive, especially when the costs of energy, collection and transport are considered.

Certain self-correcting mechanisms constantly affect the nature of resource 'depletion'. When a resource becomes relatively scarce in the short term, market forces drive up its price, which results in its conservation. This may provide an incentive for more exploration, which may in turn lead to new discoveries, and an increase in the prices of products made from that particular resource. This normally reduces public demand for those products and usually reduces the demand for the resource. The price increase on the temporarily scarce resource can also make the price of other resources relatively cheaper, and encourage the use of substitutes. In many cases new technology is developed to reduce this reliance or replace the resource. The theory is that we can reuse rather than do without a product while lessening the strain on non-renewable resources.

The economics of recycling are concerned with diverting reusable waste from landfill, thereby lessening the costs of landfill operations and garbage collection. Theoretically, recycling could pay for itself if the collected materials were sold for other uses. In some cases this has worked remarkably well, especially for companies such as Pratt Industries and Simsmetal, which have special factories using recycled materials to create products for which there is strong consumer demand. However, when materials are in over-supply, or prices have plummetted (as in the case of aluminium cans) it becomes much more costly to maintain a comprehensive recycling operation than to dump waste in landfill.

Recycling is normally much more successful when primary resources are scarce. In Australia it is a more complex issue, since this country is high on resources and low on population. Even though recycling may alleviate disposal problems and enhance the environment, until it is economically viable it will not be taken up by many industries.

One of the most visible forms of waste is packaging waste. The Industry Commission states in its 1991 report on recycling that household waste makes up about 30 per cent of the waste going to landfill, and about one third of that household waste (about 10 per cent of the landfill total) was packaging material. The environmental impact of packaging has come under much scrutiny in the last decade. Gone are the days when everything was indiscriminately discarded. However, a few things must be qualified here. Packaging in itself is not all bad. It performs important functions such as keeping food from spoiling and protecting products from breakage. Over the past decade many forms of packaging have been changed so that they either use less energy and materials in their manufacture and/or are recyclable. Also, assessing environmental impact involves not just the amount of waste produced but the vital question of where and how it is disposed of.

5.2

Making light of packaging

A CASE HISTORY FROM ACI PETALITE

There have been major advances in reprocessing the plastics used in packaging. In 1982, three years after ACI Petalite began operations, the first post-consumer PET bottles were collected from kerbsides.

The reuse of products is still a developing area. The recycling of plastics, like all recycling programs, depends on separation at the source. Plastic is a relatively low-energy raw material and this is reflected in its value for reuse purposes.

Based on present Australian kerbside programs, over 50 per cent of plastic waste could be collected and recycled. However, the issue is complicated by the fact that not all plastics are compatible when broken down. Until the infrastructure for recycling plastics is integrated into daily life, and there are greater economic incentives, in many situations it will not be cost-effective.

ACI Petalite expanded and supported the retrieval programs first established by ACI Glass and today some 4 million homes regularly recycle PET and glass through kerbside programs. In some areas the PET recycling rate is exceeding the Federal Government's 25 per cent national recycling target for rigid plastic containers, which must be achieved by 1995.

Markets for recycled PET from the recycling facility at Wodonga were established early, both in Australia and overseas. In a world first, non-food bottles were made from 100 per cent recycled PET and launched in November 1991, and recycled PET in flake form was exported for the manufacture of carpet. Since the inception of its recycling program, ACI Petalite has progressively put both the 'hardware' and the markets in place, with the centrepiece being the southern hemisphere's largest

PET recycling plant, opened in Wodonga in 1991. Last year the plant recycled 100 million bottles.

PET recycling involves a hierarchy of end-users and for ACI Petalite the replacement of virgin resin in food-grade applications was always the principle objective. In order to move forward on this, ACI Petalite and Continental PET Technologies US, both members of the BTR Nylex Limited Group, joined forces to develop a multilayer Repete (a registered trademark of ACI Petalite) system. The technology that allows the production of multilayer preforms—the first step in PET bottle manufacture and the stage where the recycled content is introduced—was the result of new sequential co-injection moulding techniques. The innovative system has been approved by both American and Australian health authorities.

The multilayer Repete system comprises a process that laminates post-consumer recycled PET between an inner and outer layer of virgin PET in the wall of the container. This provides a functional barrier between the recycled PET component and the contents of the bottle, thereby preventing the migration of possible contaminants. Multilayer Repete will have a major impact on Australia's PET recycling rate since ACI Petalite's beverage container will contain 30 per cent recycled material. There is further potential to increase this recycled content to 40 per cent, and this is expected to lead to a doubling of Australia's recovery rate in the medium-term. The high volume, end-use markets drive recycling, and the multilayer Repete system now guarantees that end use.

5.3

Improving the environmental performance of plastic

A CASE HISTORY FROM KEMCOR

Kemcor is the leader in the polyethylene field with a 40 per cent market share, and the only company in Australia capable of manufacturing the full range of polyethylene resins, including high density (HDPE), medium density (MDPE), and low density (LDPE). The Melbourne-based company, which started out in 1961, is now owned jointly by Esso and Mobil.

In its brief history Kemcor has developed a streamlined, efficient manufacturing complex producing the feedstock for polyethylene production: the resin that produces familiar domestic items such as Glad bags and film, and polyethylene bags. It also makes resin for micro-irrigation pipes, moulded plastics, plastic milk bottles, and communication cable conduit.

Plastics play an integral part in just about every area of our lives. They provide inexpensive, strong, hygienic packaging; durable building materials; household products and appliances; clothing and footwear; toys; and medical supplies. Plastic products are lightweight and, in comparison with many other materials, require less energy to transport. Plastic can also be designed for recycling by minimising the number of plastics in one product, or at least providing clearly identified break points for ease of separation into similar material; by avoiding dissimilar adhesives and paper labels; by using colorants sparingly; and by including international plastic code markings. Despite this, plastics are perceived to be environmentally unfriendly and a major contributor to domestic waste. They are made from non-renewable resources and they are not readily biodegradable. Yet an audit released by the Litter and Recycling Research Association (Victoria) in November 1993 found that domestic waste contained less than 4 per cent plastic, not including the plastic that has gone back into the recycling system.

It is important to realise that many plastic products play a useful role in our lives. The question is not so much how to get rid of plastic, but how to ensure that the manufacture, consumption and disposal of plastics is done wisely. Australian manufacturers of polyethylene resins and plastic products are seriously tackling their environmental impact, from the processing of materials through to the disposal of waste products. Members of the industry believe that if you remove the risk of pollution then you avoid the problem of having to fix it. One of the practical steps in this strategy is an environmental audit. This means that the company assesses production methods in all areas and stages of its operations from the raw materials to disposal and waste treatment, and right down to packaging and shipping.

Cynics might say that this assessment strategy is fear-driven since environmental issues have to be factored into the business equation to avoid the repercussions of ignoring regulatory guidelines and legal requirements. In New South Wales and Victoria, for example, company directors may be personally liable if pollution is attributed to their failure to exercise 'due diligence'. However, Kemcor sees the situation differently. To be credible as a good corporate citizen means being environmentally responsible. Responding to environmental issues is no longer an option, it is the way to conduct everyday business. In our case an environmental audit means a commonsense look at our planning technology, general housekeeping, maximum use of resources and sophisticated waste management. Some of the best input has come from our plant supervisors, who are closest to the day-to-day operations.

Kemcor is one of the five member companies of the Altona Chemical Complex, Australia's largest centre for the production of plastics, petrochemicals and synthetic rubber. The complex is spread over 207 hectares. The Altona community has had a very significant impact on Kemcor's environmental strategies. A high percentage of its workforce lives in the western suburbs, and Kemcor has taken an active role in working with local community groups such as the Altona Complex Neighbourhood Consultative Group. The relationship between the chemical companies and local residents has been tumultuous, but mutual understanding and trust are now well established.

Kemcor's strategies include a commitment to reducing all volatile organic compounds (VOC) emissions from their Altona plants by 50 per cent over a 5-year period. Kemcor has also taken an active role in the Australian Chemical Industry's 'Responsible Care' Program, which encompasses all aspects of the manufacturing process from transportation, emergency response and community awareness, to waste management, operational safety, and responsible management of chemicals. Although it will take some years for all the member companies to comply with the Codes of Practice, the requirements are more stringent and comprehensive than any current Australian legislation.

Environmental benefits can also be economically sound. Kemcor has initiated several environmental strategies that have already achieved substantial results. For example, some small modifications to equipment, along with procedural changes, have made it possible to reduce and recycle waste rubber from rubber drying and baling processes; and bearing in mind that one person's rubbish is another person's treasure, Kemcor has initiated waste rubber sales for low-grade purposes such as rubber tiles and conveyer belts as part of its business plan. Initiatives such as these have led to a reduction in total waste rubber from 570 tonnes in 1989 to 149 tonnes in 1991.

Kemcor has also adopted some enterprising strategies for recycling and reducing the amount of material entering the plant's wastewater system. For example, the water drains on several oil tanks have been rearranged to reduce substantially the amount of oil entering the wastewater system. The tanks are routinely drained for water removal. Compressor oil is no longer drained to the wastewater system when the oil filters require changing, but is now pressured back to the oil tank. Closed loop sampling, where only minimal material is taken from the process for testing, has allowed a great reduction in the quantity of material drained to the system.

Odour emissions have also been successfully reduced. Odour in the local community had been an issue, due partly to the closeness of residential developments to chemical plant operations thanks to inappropriate land use planning in the past. One improvement was the installation of a high-density polyethylene cover over two wastewater treatment ponds to prevent odorous emissions. A leak control program was also introduced to reduce hydrocarbon emissions. This program regularly monitors potential leaks with a hydrocarbon 'sniffing' instrument. Leak sources, such as valves, are immediately rectified or later repaired or replaced. Computer records are kept for all 20 000 valves and other items of equipment in the system. The percentage of leaking valves has been reduced from 10 to 2, with a corresponding reduction in waste emission of several hundred tonnes of gas per year.

Significant reductions in the emission of benzene to the atmosphere have been achieved over recent years. Apart from the leak control program, a thermal oxidizer and enhanced gas collection system in the rubber dryer building and the gradual substitution of benzene with cyclohexane as a process solvent for the manufacture of butadiene rubber (cleaner production) have reduced the benzene emissions by 80 per cent since 1989.

The first energy crisis in 1973 set the stage for what is happening now. A US government official once quipped 'The Shah of Iran is the plastic industry's best friend'; the price of oil doubled and it put the value of plastics in perspective. Companies began looking at recycling—at conserving the resource and gaining the maximum benefits from hydrocarbons. The technology that we see today was developed in the 1970s in response to the crisis. This includes product design, economics, industry infrastructure, efficiency, management, training and waste management, as well as the design of new polymer formulations. Until then many plastics manufacturers believed that plastic recycling would never become an issue. According to one manager 'We argued that we were doing our bit for the environment. By using plastic we were significantly reducing the energy and waste generated by other forms of packaging.'

Kemcor's shareholders, Exxon, Mobil and Woodrose, own Full Cycle Plastics Pty Ltd, a plastics recycling company in Cheltenham, Victoria. The plant reprocesses high density polyethylene (HDPE) milk bottles which are collected in all mainland states. The material is chopped up and washed, impurities are removed and the remaining material is extruded into beads almost identical to new HDPE. About 100 tonnes per month of material are being recycled and this is steadily increasing. Plant capacity is about 3000 tonnes per year and total sales of dairy resin are about 18 000 tonnes per year.

The key to an increase in throughput is to find extra markets for products containing recycled resin. Products that can replace steel, timber or concrete may also be beneficial by reducing other resources the community consumes. The recycled HDPE product is sold for the manufacture of crates, forklift pallets, bottles and pipes. More and more councils have established bottle collections, and some have shown interest in using the recycled bottles to make signposts, wheely bins and other useful items. This is to be encouraged as increasing demand for the recycled resin will enhance the success of the full recycling venture.

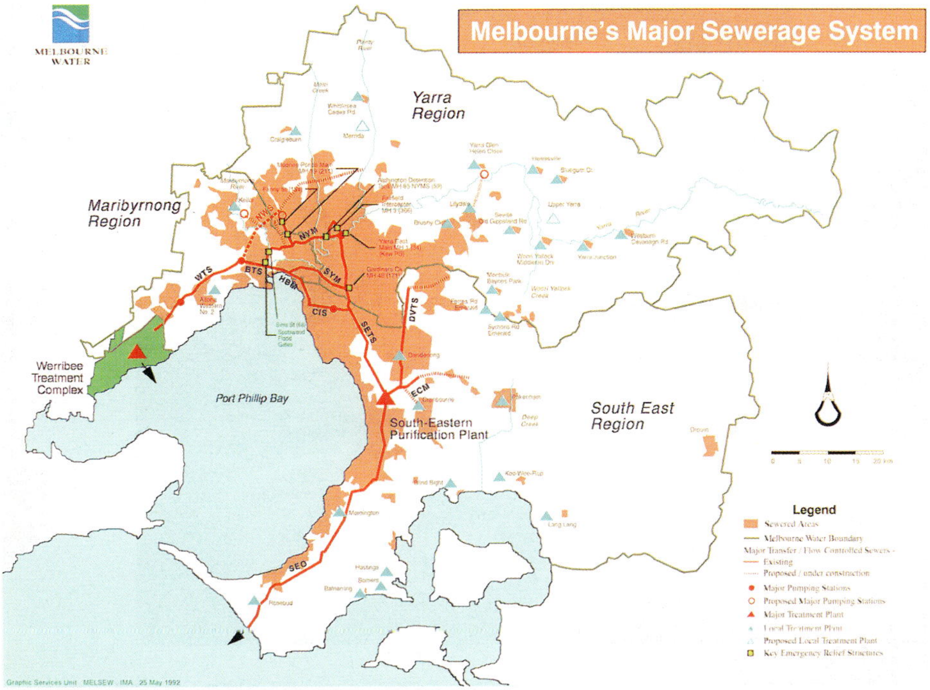

Above Melbourne's major sewerage system. *Right* Flowers grown with the aid of treated sewage from SEPP.

Above Examples of different containers made from liquidpaperboard. *Below* Bales of used liquidpaperboard packaging collected and ready for recycling.

Above Tile sorting and recycling. *Below* Halon Fire Extinguishers for recycling at Dascem's Halon Bank.

Batters of the old Yallourn north extension open-cut mine in the Latrobe Valley *(above)* before rehabilitation and *(below)* after rehabilitation.

5.4

Infinitely recyclable

A CASE HISTORY FROM ACI GLASS PACKAGING

Glass has been around for about 4000 years but recycling it on a large scale is a recent innovation. ACI has been recycling glass for more than 100 years—since it began operations in Australia. Since 1986 glass collection programs have been upgraded substantially, and ACI has taken a lead in assisting local government to establish and promote kerbside collections throughout the nation.

As overall community and political awareness of the recycling challenge has grown, so too has the level of support from state governments, local councils and householders. In Australia, kerbside collection schemes, drop-off centres (igloos and bottle banks) and bottle depots are the most usual methods of collecting glass for recycling. Kerbside collections are usually made weekly, with each household using a recycling container (bag, crate or bin) for jars, bottles and other recyclable glass products. A recyclables collector empties the container at the kerbside and takes the items back for processing.

There are several advantages to kerbside collection. Householders do not need to use their own transport since recyclable items have to be taken only as far as the kerb. The regular, weekly collection day is easy to remember. Councils pay for the service and the fee may be offset by the savings on waste disposal. The scheme also helps to educate the community about recycling and waste reduction and those participating influence others by example. If it becomes more economical to recycle, it may be possible to include other products in collection services.

The most common alternative to kerbside glass collection is to provide igloos and bottle banks at drop-off centres in easily accessible places such as supermarket carparks, service stations, scout halls or local tips. The drawback is that consumers must use their own transport, which also means more fuel consumption and emissions. Drop-off centres are usually used where population density is too low for kerbside collections. In high-density areas the two systems may complement one another.

In Victoria 100 per cent of metropolitan households and over 90 per cent of rural households are served by a weekly kerbside recycling collection. In New South Wales 98 per cent of Sydney households have a kerbside collection service and this is expanding to country areas. In Queensland the first kerbside collection services began in 1991, and now there are collections in Brisbane and most of the other provincial cities. Before kerbside collection was introduced, Queensland had an extensive drop-off network. The Northern Territory has drop-off centres in Alice Springs, Katherine, Darwin, Tennant Creek and many smaller communities.

In Western Australia Perth has a kerbside collection for most households and over 100 igloos and there are drop-off facilities throughout the state. South Australia has relied on legislation and 'forced' deposits to handle the collection of beer and soft drink containers. This is the only state that has preferred to legislate for deposits rather than rely on voluntary, comprehensive kerbside and drop-off systems. In Tasmania drop-off centres are the most common method of glass collection. Kerbside collection is gradually being introduced to more populated areas. The success of kerbside collecting with a complementary drop-off network is illustrated by the 147 per cent increase in the amount of glass recycled nationally between 1987 and 1993.

Glass recycling rates vary from state to state. Most are improving rapidly. The factors influencing each state's recycling performance include population density, urbanisation, collection methods (or the extent of weekly kerbside collection), legislation, public education, local and state government support, and community participation. The glass industry is aiming for a national 45 per cent recycling rate by the end of 1995.

The final stage of the glass recycling process is, of course, selling the recycled end-product in the marketplace. As increasing amounts of cullet (recycled glass) are used it is essential to meet tight quality standards. Contaminants such as metals, ceramics, plastics, stones and general debris must be removed to maintain the consistently high standards required. The recycled product must compete with the original in price and quality if it is to be successful. Recycled glass products are indistinguishable from glass products made from virgin raw materials and the only real limitation on the amount of glass that can be recycled is the demand for new glass products.

As well as recycling, ACI Glass Packaging has been reducing its environmental impact in other areas. For example, since 1981 energy efficiency has been improved by using natural gas as the main source of energy, by continuing to reduce the weight of glass packaging, and by improving furnace design and melting techniques. Australian-made glass now contains an average 40 per cent recycled glass. The energy savings made by using cullet (glass waste) vary depending on the efficiency of the furnace, the amount of glass being produced and the proportion of cullet in the raw batch mix. For example, glass made with 100 per cent cullet uses 20–35 per cent less energy than glass made from virgin raw materials.

Research into ways of manufacturing lighter glass containers has already led to a significant reduction in energy and raw material use. For example, in 1986 a 'stubby' weighed 260 grams, but now weighs about 160 grams—a 38.5 per cent reduction. Advances in light-weighting technology are continuing, since every improvement lessens the use of energy and raw materials, and reduces emissions.

Unlike public recycling programs, company programs that reduce the use of energy, virgin raw materials, emissions and solid waste are not immediately apparent to the consumer; they do, however, make a significant impact upon the environment. ACI Glass Packaging is continuing its commitment to glass recycling programs and other forms of waste minimisation.

5.5

Multiple strategies, multiple use

Association of Liquidpaperboard Carton Manufacturers

Assessing environmental impact involves not just the amount of waste produced; but where and how it is produced and disposed of. Take, for example, a litre carton of orange juice weighing just 30 grams which leaves behind at the orange juice plant about 1.2 kilograms of orange peel. Most households would dispose of that 1.2 kilos with other household garbage, unless they have a compost heap. However, the peel discarded at the orange juice plant can be sold for animal feed or compost. The argument is that a 30 gram carton can save 40 times its weight in waste that would otherwise go into landfill—cartons, like other forms of packaging, can save waste.

This is further illustrated by a study conducted by one of America's 'garbage archaeologists', Professor William Rathje, who compared the waste thrown out by American and Mexican households. Although the Mexicans used a lot less packaging, they created about 43 per cent more waste per capita than the average American householder did, since their waste mostly consisted of food scraps and fruit and vegetable peelings. Packaging, on the other hand, distributes only the useable or edible part of food. It also preserves food, and in some kinds of packaging (e.g.) 'long life' cartons) certain food no longer requires refrigeration. The more efficiently a product can be transported, the fewer the transport requirements. Producing food in environmentally efficient packaging benefits the environment.

What makes packaging environmentally efficient varies according to the circumstances. In some cases a milk carton may be a better choice than a milk bottle. Much depends on the location, population size and local facilities. One difference between a refillable glass bottle and a 'one-trip' milk carton is weight. The bottle is heavier, it uses more resources in its manufacture, and its transport requires more energy. On the other hand, a milk carton is light, weighing only 3 per cent of the weight of the milk it contains. The carton not only uses less material, but reduces the energy required to transport the milk, and it takes up less landfill space than a glass bottle does.

Like every complex problem there is no simple answer for waste and it requires several strategies: legislation, education, and a combination of recycling, composting, converting waste to energy and source reduction. All of these incur costs. Some people maintain that if the environmental cost of packaging was included in the price of the product, consumers faced with the real costs of packaging would use less.

The cost of recycling has been higher than expected in many cases and it has been suggested that packaging manufacturers assume responsibility for recycling their products. Many manufacturers think that this method of funding recycling is inefficient and inflationary, and would ultimately save very little landfill space. They believe that if recycling is not supported by local government, packaging costs will increase and this could make some Australian companies uncompetitive.

There are other consequences. In the 1993 study, *The Economics of Packaging and the Environment,* Dr John Hatch and Dr Trevor Mules demonstrate with economic modelling that any tax or levy on packaging, including the cost of imposing an arbitrary recycling target, would depress the use of packaging. The environment would suffer as a result.

There are those who believe that much of the waste going to landfill should be burned and the resulting energy harnessed to render the energy content of this waste 'recyclable'. Converting waste to energy offers many advantages. Whereas a relatively small proportion of collected waste is recyclable, and some 50 per cent is compostable, nearly 90 per cent of waste is combustible. Burning waste significantly reduces its volume and removes the health hazard of rotting garbage. The remainder is normally used as clean fill or road-base.

Modern waste-to-energy (WTE) plants have sophisticated emission control systems to meet stringent government standards. By using garbage as the energy source, they help to conserve fossil fuels. Those running WTE plants say that garbage burns more cleanly because it has a lower sulphur content. Singapore burns about 85 per cent of its waste. In Denmark, the plant that services the cities of Copenhagen and Fredricksberg converts to energy about 300 000 tonnes of refuse annually, generating 425 000 MWh of electricity. Three such plants in the greater metropolitan area of Paris service 10 000 000 people, generating energy as electricity and steam to heat homes.

Liquidpaperboard is a recyclable material. However, some liquidpaperboard products are being used by the Swiss and the Germans for fuel, and Australia could consider this alternative to recycling. Cartons are already used here as tree planters by organisations such as Greening Australia and as a craft source in pre-schools. Cartons are being extensively recycled in Australia to produce high-quality office paper.

Whenever you recycle paper some of it is always wasted because the small fibres it contains become unusable. These are separated out, but after about five cycles a certain fibre becomes unusable in paper manufacture. We have begun to investigate other uses for this resource and the obvious choice is to extract its energy potential, especially for remote areas without recycling facilities. Cartons work well as fuel. They are clean, high in energy, and burn readily, with very few emissions except carbon dioxide. This is an example of a 'biofuel'—a fuel that originates from biological systems like trees. The carbon dioxide emitted in burning is balanced by the carbon dioxide that is absorbed in growing the biofuel. When cartons are burned the carbon dioxide load is greenhouse-neutral.

Apart from its recycling advantage, liquidpaperboard packaging offers other environmental benefits. The cartons are made from a renewable resource—trees that are harvested from managed, sustainable forests. The cartons are degradable when cut up and can be added to compost. ALC employs a full-time environmental affairs manager to promote the recycling possibilities of Liquidpaperboard products. It promotes recycling in Australia in two ways—through local councils (of which there are 850) and by mass communication media and marketing directed at the general public and schools. Often a recycling message is printed on the carton.

Ultimately the selection of packaging and other community waste issues will need an integrated approach involving source reduction, composting and converting waste to energy, as well as recycling. Fortunately all those strategies can be applied to cartons: they reduce waste at the source because of their light weight, they are recyclable and compostable, and can be safely burned to recover energy. They also offer several reuse options, of which their use as tree planters is the most interesting.

SECTION 6

Waste not, want not

6.1

Waste management accountability

LEIGH CATO

While consumers continue to claim that companies and organisations are ruining the environment, population increases mean that individuals must also be accountable for their impact on the environment. Accountability must be public as well as corporate.

In the past twenty years there have been many changes in what people dispose of through the sewage system. For example, recent health concerns and social changes have led to an enormous increase in the number of condoms passing through the sewage system; it seems to be the most discreet way of disposal. Condoms, sanitary napkins and disposable nappies with plastic liners have all put a strain on conventional sewage treatment. However, the most diabolical item of all to enter our sewers is the humble cotton-tipped stick ('cotton bud'), once made of cardboard or wood, which breaks down, but now mostly made of plastic. Many cotton buds escape screening at the treatment plant and end up on our beaches.

Beachcombers looking for shells now find our shoreline littered with pink plastic strips which are not only unsightly but pollute our waterways and oceans. Who should be held responsible? Government, for insufficient legislation? Manufacturers, for making these products? Users, for not disposing of them responsibly? Or those in waste management, upon whom the problem is literally dumped?

6.2

The history of environmental management in Victoria

NORM PARRIS, MANAGER INDUSTRY SERVICES,
ENVIRONMENT PROTECTION AUTHORITY, VICTORIA

According to the Environment Protection Authority (EPA), whilst it is essential to have efficient systems to manage waste, improved environmental performance depends on developing programs that prevent or reduce it. They are convinced that cleaner production can benefit industry, the community and the environment. Many of the steps that firms can take towards cleaner production result in short payback periods. Relatively straightforward changes to housekeeping can deliver dramatic improvements. Investment in cleaner production can not only pay dividends for business and the community, but can deliver these benefits in the long term as well as the short term.

For several years the EPA has realised that significant improvements in environmental performance would not be possible if it concentrated solely on the discharge end of the pipe, and so the EPA has explored opportunities to prevent pollution by examining the earlier stages of the manufacturing process. Its experience in this area has revealed the enormous gains that can be made by taking a fresh, innovative approach.

The importance of environmental management is now at a point where it is an integral component of the general community's value system. The concept of 'sustainable development' and the necessity to manage the environment more efficiently for the future are key issues. Waste management is a particularly important issue in the broad spectrum of environmental management.

The community has already embraced a range of environmental issues, which range from so-called intractable waste to personal recycling. The generation and disposal of the waste and by-products of our industrialised society must therefore incorporate several factors including sustainable development, technology, cleaner production and waste minimisation, and community concerns.

The fundamental challenge for the community, and that includes manufacturing and the consumers of its products, is to develop a comprehensive approach to environmental management that is not isolated from the other factors that shape our behaviour and decision-making. The pollution of our urban and rural environments should not be separated from product manufacture, selection, use, reuse, recycling and disposal.

The Brundtland Commission popularised the term 'sustainable development' in its report 'Our common future'. Although there is a general understanding of the term, there is less agreement on what specific programs would achieve its objectives. The Inter-governmental Agreement on the Environment, finalised in February 1992, placed sustainable development in the Australian context and one of the points it makes is that environment and development are two sides of the same coin and are integral components of the decision-making process; the costs of environmental controls must be consistent with the benefits they aim to deliver.

One of the principles of the Agreement is the often misquoted 'precautionary principle', which in effect means that decisions should be approached with caution when there is uncertainty about intergenerational equity, biodiversity and ecological integrity. In simple terms this means that irreversible damage should be avoided and the ecosystem retained for future generations. In this context, environmental protection is a far broader issue than the control of waste emissions to specific standards; it requires closing the loop between community expectations, development decisions and their resource and disposal implications.

In the early 1970s the environment protection program faced a pollution load that was the legacy of a good many years of unco-ordinated and inadequate action. The early years of EPA's development focused on providing a policy framework for control strategies. Undoubtedly many local improvements were made, but in many respects this era tended to focus on dilution as a means of improving environmental performance.

Whilst the concentration of pollutants remains important in some respects, by itself this method could not deliver the load reduction required for a long-term improvement of the urban environment. Consequently, pollution control technology received greater attention. To meet statutory policy objectives in air, water and land environments, many industries wishing to discharge their wastes installed equipment solely to control emissions. However, in the process they added very little to the productive capacity of the plant. Again, significant gains were made in environmental protection but in the long term the law of diminishing returns would constrain initiatives. Although the control of major sources of pollution improved the air and water environments for the community, proportionally diffuse sources became more significant. Also, whilst the transmedial nature of environmental management was certainly recognised by experts, it was not generally appreciated by those in the technical disciplines dealing with the issues. The additional capital and operating costs of pollution control equipment added to the popular depiction of environmental protection and economic development as two conflicting and mutually exclusive regimes.

With growing awareness of global environmental issues and the popularisation of several terms—including 'sustainable development'—this decade could be depicted more accurately as one where the focus shifted towards preventive strategies. Although the environmental outcomes of this attitudinal change can deliver far greater rewards than are possible through the traditional command and control strategies, economic considerations are equally important. Certain programs have been developed that emphasise the holistic nature of waste and environmental management. Rather than become distracted by semantic discussions on the origins of these initiatives, it is important to recognise their fundamental similarities. If waste is regarded as a misplaced product, it is obvious that a range of disciplines, operating personnel and management have a keen interest in minimising costs by maximising production yields. Not only does waste generation represent an opportunity cost, but the sting in the tail arrives with the account for its treatment and disposal.

The shift towards pollution prevention strategies has come from several directions and includes programs generalised by terms such as cleaner production; waste minimisation; best-practice environmental management; best-practice environmental regulation; total quality management; and pollution prevention. These strategies include a cradle-to-grave concept for managing waste. By requiring waste generators to contribute to the proper management of the material even after its departure from their premises, these strategies provided the all-important basis for subsequent pollution prevention programs. The provision of suitable transport, treatment and disposal facilities highlighted the real costs of the generation of industrial waste and provided incentives for reducing it.

Three major components—legislation, facilities, and support services—were identified. Access to support services (in the form of information on cleaner technology and processes), training programs and community involvement provided the natural link with cleaner production strategies. Traditional approaches to waste management have tended to focus on end-of-pipe solutions, usually involving burning or burying. The more recent moves toward waste minimisation demand useful alternatives and disposal by reducing waste at its source. Waste prevention must begin at the initial planning stage of any new product or in the siting of new operations and may require different raw materials to manufacture the product, or in the extreme, rule out manufacturing the product altogether. A notable example is the phasing out of chlorofluorocarbons.

Total prevention is seldom realistic and waste reduction becomes the aim. Consistent with the waste minimisation hierarchy, there are several ways to reduce waste, such as adopting new production processes that generate less waste; modifying production processes; and recovering and recycling waste process materials.

CLEANER PRODUCTION PROGRAMS

The EPA in Victoria has set up several programs that have helped firms in key industries to recognise the value of cleaner production and to change their technology and practice. These changes have led to a significant shift away from dependence on end-of-pipe technology. Of note here is the Cleaner Production Grants Scheme where the government has financially assisted Victorian firms to introduce innovative cleaner production technology in the state. The program is aimed at supporting new technology or processes that minimise the generation of waste rather than merely treating it in a more efficient manner.

One of the various industries supported by the scheme is Quality Heat, a firm that replaced molten cyanide salt baths for hardening metal with a fluidised bed technology that uses alumina. Alumina behaves like a fluid at high temperature and simulates the heat transfer characteristics of the cyanide salt. The process produces a superior product but completely avoids the environmental and occupational hazards of the technology it replaces, and offers considerable energy savings as a further incentive for its introduction.

Another example of cleaner production comes from the farm sector. Twelve thousand pigs produce as much waste as 40 000 humans. A Ballarat piggery installed a methane digester plant to produce biogas. This generates enough electricity to run the farm and a 50 per cent surplus that is sold to the electricity grid, and the digester sludge can be used in compost. This plant converts piggery effluent, with a high pollution potential and treatment cost, to an asset that brings an acceptable return on investment.

The EPA, the Department of Business and Employment, and several leading firms worked together on a cleaner production demonstration program that illustrates the value of incorporating pollution prevention in the manufacturing process. Firms in the program were chosen according to several criteria, not the least important being a good track record in environmental performance and prominence in key industry sectors. The seven firms in the program are Cadbury-Schweppes, Ford Motor Company of Australia Limited, WR Grace, Conagra Wool Pty Ltd, ACI, BASF and Smorgans. Features of the program include: management commitment to the program; staff involvement in its execution; EPA and DBE assistance with waste minimisation assessments; and, most important, the incorporation of waste prevention thinking into the firm's day-to-day operation. Waste minimisation must be regarded as a continuous quality management program by everyone in the enterprise.

Early results are very encouraging. Cadbury-Schweppes, for example, has identified significant production savings at no cost, and an additional 14 projects with much greater annual savings. All have payback times of less than two years; many are considerably less than this. Another program is the Melbourne Water Industrial Waste Grants Scheme that was set up in 1986 to encourage high-quality projects in waste minimisation and recycling, particularly in the field of industrial waste. The scheme has funded such projects as wastewater reuse in dye works, the development of water-based adhesive substitutes, tannery waste minimisation, and the minimisation of volatile organic emissions.

An Industrial Waste Grant helped Oiltech Australia develop a mobile oil recycling unit that recycles onsite certain types of industrial oils without interrupting production processes. Some 1.5 million litres of oils and lubricants could be recycled each year by the mobile unit with considerable cost savings and waste reduction. About 60 per cent of industrial oils and lubricants have previously been disposed of as waste. The facility can also be used by municipal councils for recycling specific grades of collected oils and lubricants.

The Colloid Centre Research Group, led by Professor David Mainwaring at Swinbourne Institute in Victoria, was awarded a grant for research on foam and flotation technology for treating industrial wastewater effluents. A further grant was awarded to incorporate the promising results to a large continuous operating system. The process reduces heavy metal contaminants and other wastes from plating factories, and overcomes two major problems with the conventional metal waste treatment: the high alkalinity of effluent, and large volumes of wet sludge. A wetting agent or surfactant is added to the waste and foam, produced by air injection, collects the heavy metal hydroxide particles and floats them out, leaving a high-quality effluent suitable for recycling.

Cleaner production has been espoused by the EPA for years. If it is to become thoroughly accepted by industry, it must be more closely structured to meet the needs of industry. To help achieve these objectives, the EPA has established the Australian Centre for Cleaner Production. Although initial administration and seed funding was provided by the EPA, its development will be overseen by a board comprising representatives from industry, tertiary institutions, research interests and government. The centre will liaise between industry, government and academic circles to help co-ordinate research on cleaner production. It will also set up demonstration projects in industry and offer training courses and plant assistance with waste minimisation assessments. The centre will link related programs in Australia and overseas, enhance links with the United Nations Environment Program (UNEP) cleaner production program, and explore bilateral links with programs in neighbouring countries.

PRODUCT DESIGN: THE FIRST STEP IN WASTE MANAGEMENT

The proposition of sustainable development raises the issue of resource use in terms of better product design. Much of the cost of a product's manufacture is incurred at the design stage, so it is not surprising that companies find it inappropriate to pass a design to the manufacturing phase without first involving production engineers. It is also acknowledged that design should incorporate environmental objectives, and this is becoming easier to do without inhibiting product performance or usefulness.

The EPA believes that since a waste material is a 'displaced product', a product should be designed for waste prevention to minimise costs. However, designers rarely think about how their products will be managed after their useful life is over, and waste managers tend to develop systems to handle the prevailing waste streams. Extended to product design, cleaner production involves closely relating each stage of the life cycle to its constituent resources. Examples of this approach include the design for disassembly as well as assembly; products and components that can be re-manufactured; and better materials management to retain the value of a resource. In some cases this means redesigning to reduce the toxicity of materials and by-products, and incorporating energy recovery into the waste management regime.

Waste has been an inevitable by-product of our daily lives. However, as global concerns about our environment grow, communities and industries alike are under pressure to manage their waste better. The public's growing awareness of the need to prevent waste extends to hazardous materials and solid discards such as paper and plastics. Green design will become increasingly prominent in our manufacturing, purchasing and disposal decisions. The EPA supports innovative approaches to environmental management and has set up programs that emphasise working with industry to incorporate waste reduction or cleaner production into its practices.

6.3

World leadership in wastewater technology

Case histories from Melbourne Water Corporation

Melbourne Water Corporation was formed in 1991 after the Melbourne and Metropolitan Board of Works amalgamated with several suburban water authorities. Melbourne Water is in the interesting, if not enviable, position of receiving the largest concentration of industrial waste and domestic waste in the country. It is also a world expert on wastewater treatment. The Western Treatment Plant treats about 50 per cent of Melbourne's wastewater and the Eastern Treatment Plant treats about 40 per cent of it. The remainder is treated by local plants.

Environmental management system

In the past 60 years not only has Melbourne's population more than doubled, but the size of the metropolitan area has increased sixfold and the number of rateable properties fourfold. Per capita use of water, the generation of wastewater and run-off volumes into the stormwater system have also doubled. For much of that 60 years Melbourne Water's main targets have involved quantity—supplying enough water to avoid droughts, cope with wastewater flows and stave off flooding from stormwater runoff.

However, the challenges Melbourne Water faces now and in the future are to control quantity problems through demand management, to meet environmental requirements for wastewater discharge, maintain a uniformly high standard for water supply; and develop strategies to improve the quality of stormwater runoff and thus prevent litter and other pollutants finding their way into Melbourne's rivers, creeks and bays. To meet these challenges and responsibilities, Melbourne Water has developed an environmental management framework consisting of eight major programs. They include an environmental policy, management plans, assessment, employee training, community involvement, monitoring and research, audits, and performance reporting and benchmarking. For Melbourne Water to achieve its goals, households will also have to take part in dealing with waste management and the impact of waste on the environment.

WASTE MINIMISATION

In the treatment of wastewater at the Eastern Treatment Plant (ETP), Carrum, as many 'natural' processes as possible are used. In the 'activated sludge' process, naturally occurring micro-organisms are used to feed vigorously on the organic material in wastewater. Primary treatment involves the sedimentation and removal of solids as sludge from the wastewater in large tanks. Secondary treatment breaks down the suspended solids and dissolved organic matter. Sludge is then further broken down by anaerobic digestion, converting 45–50 per cent of the volatile solids to liquid and gas. An eventual third stage of treatment, to remove the nitrates, would probably cost more than the primary and secondary treatments combined.

Waste minimisation is part of Melbourne Water's everyday business. For example, energy accounts for about 25 per cent of ETP's running costs. The plant substantially cuts costs by using power produced onsite during peak tariff times. Treated wastewater is also used for lavatory flushing, plant cooling and watering garden and grassed areas at the plant.

The sludge digesters collect enough methane gas to generate 40 per cent of the plant's power needs. The generators run for treatment processes during the day, and at night the powerstation is turned off, the gas is transferred to the outfall pumping station and used to operate the pumps. Usually only one pump operates during the day to keep the level right and to let the treated wastewater build up in the holding basins. At night the basin levels are pumped down. This way, efficient energy results are achieved. Like any other business, it is a matter of economics. The aim is to produce safe, clean effluent at an acceptable cost.

The waste stream is not a constant flow. On weekdays, most people get up in the morning, have their shower and use the toilet. There is a peak flow at the plant at about 1.00 p.m. because it takes time for the wastewater to travel to it. Demand then slows down again until people get home in the evening and the plant then has another peak.

SEWAGE PRODUCTS

The Eastern Treatment Plant produces three marketable products: sludge, treated wastewater, and methane gas. Viable businesses using these by-products have been initiated. Since the plant was commissioned in 1975, it has always been Melbourne Water's policy to encourage the reuse of treated wastewater and sludge. Currently some twenty different horticultural enterprises ranging from flower growers to vineyards are tapping into the 56 kilometres of pipeline that carries 400 megalitres of treated wastewater from ETP to Bass Strait each day.

The treated wastewater nourishes golf courses, turf farms, specialty nurseries and even a cemetery. One flower grower from Rosebud, Victoria, said that tapping into the pipeline has allowed him to increase production at his 40 hectare property by 400 per cent. According to the superintendent of the Rosebud Country Club, which has used the pipeline for 16 years, the club has been able to cut fertiliser application to a third of the amount used previously. The treated wastewater is also cheaper and better for the grass.

Melbourne Water and the community have become increasingly concerned about the waste of a natural resource that is rich in nutrients. A horticultural park on the plant's 1000 hectare site has provided new business opportunities. The 250-metre buffer zone surrounding the plant has been divided into 8 hectare lots and boasts a Christmas tree farm, an advanced tree-growing enterprise, a hydroponic fruit and vegetable operation, an instant lawn farm, and a cut-flower business.

One of the plant's major products is sludge, which after seven years can be used as fertiliser. A-1 Soils, a fertiliser company near the plant, has developed a flourishing business by mixing sludge with sawdust, woodchips or other substances to produce high-quality blended soils, composted soil conditioners, and potting mix.

A NATURAL SOLUTION TO POLLUTION CONTROL

The Western Treatment Plant, halfway between Melbourne and Geelong, has been treating Melbourne's wastewater using a land-based process for over 100 years. The plant is acknowledged as a world leader in the development of environmentally friendly land/lagoon technology. It is 35 kilometres south-west of Melbourne, comprises of 10 850 hectares and treats 174 000 megalitres of wastewater annually (50 per cent of Melbourne's wastewater). It uses three treatment systems—land filtration, grass filtration, and lagooning—before the treated wastewater is discharged into Port Phillip Bay.

After raw wastewater is delivered to the plant, a network of channels distributes it to various areas for treatment. Just under half of the wastewater entering the plant (averaging 520 megalitres daily) is given pre-treatment in the form of primary sedimentation. In the case of grass filtration, pre-treatment is necessary to reduce the amount of suspended solids to manageable proportions.

From October to April, the months of high evaporation, land filtration involves irrigating permanent pasture bays with wastewater, which is generally unsedimented. Each bay is irrigated according to the prevailing circumstances. The average depth of the irrigation is about 100 millimetres with operating cycles of 14–20 days. The irrigant is treated by filtration through soil, while the contaminants and nutrients remain in the soil. About 50 per cent of the volume of water is collected as treated effluent by sub-surface drains and discharged into Port Phillip Bay. The remainder is lost through evapo-transpiration and evaporation. These richly

Land filtration

Grass filtration

Lagoon filtration

fertilised grasslands provide ideal grazing areas for a large number of sheep and cattle (up to 30 000 sheep and 16 000 cattle). This livestock generates about $3 million dollars a year which helps offset the wastewater treatment expenses.

Grass filtration involves continuously passing primary effluent through Italian rye grass on graded areas at a slow rate. The primary effluent flows onto the grass paddocks where natural bacteria removes 95 per cent of the pollutants. The grass acts as a filter, breaking down the pollutants with a bacterial film that builds up on the grass stem. Once again the effluent is collected by a network of drains and discharged into Port Phillip Bay. This treatment method is used from May to September, when there is insufficient evaporationt to allow the use of land filtration. At the end of this period the grass filtration areas are allowed to dry out, the grass then goes to seed, and cattle are grazed on the dry vegetation. The grass seed remaining on the ground is germinated by rain or irrigation in the following autumn to provide for the following operational season.

All year round, raw wastewater is channelled into the lagoons where it is broken down by bacteria, zooplankton and sunlight during detention times of up to 50–70 days. Originally each lagoon consisted of a series of 12 ponds with oxygen supplied by the natural processes of wind re-aeration and algal activity. However, to speed up the process and reduce odours, a lagoon development program has developed deeper, higher-rate anaerobic ponds with mechanical surface aeration downstream of the anaerobic reactor, and ponds for nitrogen removal have been optimised by managing the nitrification/denitrification process. An anaerobic reactor at the plant has been covered to collect gases generated by the bacterial activity of the anaerobic process. This has involved installing the largest lagoon covers in the world. By collecting and using the gas this way, not only are odorous gases eliminated and power created, but greenhouse gas emissions are decreased substantially. This initiative has been so successful that Melbourne Water plans to cover more lagoons.

Treated wastewater at the plant is also being used and tested for income-producing projects. A successful tree plantation trial has now made it possible to take advantage of the nutrient-rich water and provide a sustainable crop. After three years of growth, the average height of these trees measures about 10 metres.

Joint ventures

In a joint venture with the Melbourne University Department of Veterinary Science, Melbourne Water staff are developing a vaccine to protect cattle from beef measles, the bovine version of the human tape worm. Currently, cattle bred on land at the plant can be sold only after 18 months when there is no trace of the parasite. The vaccine would allow the sale of younger cattle, and significantly increase income.

Zooplankton is harvested from the lagoons at the plant in a joint venture, between Melbourne Water and Zootech. Harvesting zooplankton will increase income, offsetting expenses as well as sustaining zooplankton and algal activity while increasing treatment efficiency.

Wetland of international importance

One additional by-product and feature of the Western Treatment Plant is that it has been officially recognised as a 'Wetland of International Importance', for many decades. The plant is the subject of several international treaties for conserving migratory waders and other bird species and is now managed in accordance with a carefully prepared Wildlife Management Plan.

The lagoons, paddocks and mudflats of the plant are either home or summer resort to tens of thousands of birds. It is a bird-watcher's paradise, attracting birds from as far as Siberia, and it ranks with Kakadu National Park. There is no doubt that without the Western Treatment Plant this area would not be nearly as significant. The wastewater provides a fertile environment for the birds' food source—plants, zooplankton and algae.

The wetlands provide a vital refuge for the birds in the hot summer months when many of their other wetland habitats dry up. Sometimes there are 40 000—60 000 ducks or 25 000 straw-necked and sacred ibis at the plant, and it is home to about 260 species of birds. The farthest travellers are the 30 000 or so migratory waders that fly down from Siberia every summer. These tiny birds include red-necked stints, curlew sandpipers and sharp-tailed sandpipers. Some weigh no more than a matchbox, yet each year they make a round trip of more than 22 000 kilometres. The wetlands have also had a positive effect on the people who work there. Staff see themselves as not just being people who treat wastewater but also as environmental managers and caretakers.

INDUSTRIAL WASTE AGREEMENTS

The Industrial Waste Branch of Melbourne Water operates a Waste Advisory Service that provides information on water conservation, water reduction and waste disposal, and assists with negotiating and monitoring trade waste agreements. A trade waste agreement states the terms and conditions that a company must observe when discharging waste into the sewage system, and these stipulate the nature, components and characteristics of the waste; the rate of discharge; the days it can be discharged on; specific treatment and monitoring systems; and any other special terms and conditions. Monitoring is carried out by trade waste officers who visit industrial and commercial sites regularly to check the quality of the waste, take samples and make certain the treatment processes and equipment are operating correctly. Companies who do not comply with the terms of their trade waste agreement may have it terminated, which would make them ineligible to discharge into the sewage system.

Training and information services are also integral aspects of the support provided by the Branch, which mainains a flow of information and educational tools to schools, tertiary institutions, the community, and industry.

THE GRANTS SCHEME

An Industrial Waste Grants Scheme is also administered by the Industrial Waste Branch, to encourage high-quality projects in waste reduction and recycling, especially in industrial waste. Grants are assigned to specific developments and PhD projects and are available to industry, academic institutions, research organisations, government agencies and community organisations.

SEWER SENTINEL

To assist the Branch in its day-to-day operations, Melbourne Water and the CSIRO have developed a new instrument to detect companies who have not yet learned how to 'do the right thing' and who continue to dump inappropriate substances into sewers. The instrument, dubbed the 'Sewer Sentinel', continuously monitors wastewater for a range of chemically related characteristics. Wastewater is pumped past measuring probes which are specially adapted to withstand fouling in the harsh sewer environment and which can trigger alarms or automatically take samples for computer analysis. Armed with data provided by the 'Sewer Sentinel', trade waste officers can now trace illegal discharges that would otherwise go undetected. This is an important factor in making sewers a safer workplace for operations and maintenance staff.

EXPORTING TECHNOLOGY

In response to a growing national and international interest in the innovative technologies and outstanding capabilities of this organisation in managing water, wastewater, drainage and urban open space, Melbourne Water formed an International Operations group in 1992. The purpose of the group is to help the organisation grow and establish itself as a world leader in the water industry by promoting its 'ecologically sustainable development' philosophy and technology. This is to be achieved by winning substantial international contracts involving Australian industry and government.

The group's initial focus has been on intensive marketing and Melbourne Water's position in the international marketplace. Overseas countries have shown keen interest in the environmental performance of Melbourne Water, in the total systems planning capabilities, and in the management skills involved in an integrated system. In particular, Melbourne Water is marketing its operational skills in water, wastewater and environmental areas, especially innovations in lagoon systems. Proposals and submissions for work, primarily in South-East Asia, have helped establish networks with Australian government departments and agencies, strategic alliance partners in private industry, and regional government and water authorities.

Substantial progress has been made by Melbourne Water with projects completed or underway in Bangkok, Papua New Guinea, New Zealand, Hong Kong, Nauru and Pakistan. The opportunity of substantial work in West Java, Indonesia has recently been confirmed with the signing of a Memorandum of Understanding with the government and thirteen water authorities there. Melbourne Water has also been shortlisted for several major projects in China.

6.4

Cleaning up chemical cleaning

A case history from Applied Chemicals International Pty Ltd

The industrial cleaning business is significantly changing its base technology because of environmental, health and safety restrictions on materials that have been in common use for decades. Ozone-depleting substances such as 1,1,1-trichloroethane (111) and CFC 113, and those suspected of carcinogenic properties, such as methylene chloride, perchloroethylene and trichloroethylene, have led to a fundamental review of cleaning practices in the metal cleaning and electronics industries where these materials are commonly used. The Australian-based Applied Chemicals International is a leader in helping create international guidelines and standards for the chemical industry. For example, the company has instigated a replacement program in which a near-neutral water-based cleaner is used as an alternative to 111 for a manufacturer of pump valves.

Perhaps the most widely used metal cleaner in the last 20 years has been 111. Although 111 has replaced chemicals that were even more toxic, it is still not environmentally acceptable. An Applied Chemicals client, a pump valve manufacturer, was aware of the environmental implications of using the chlorinated solvent, but also required a cost-effective system. According to Ian Hicks, Chairman of Applied Chemicals, the company created an aqueous system that cost about 50 per cent less and did not reduce the quality of the finished work. The aqueous system consists of three stages: the wash cycle in cleaning fluid, a rinse cycle, and drying. Recently, vacuum drying has been introduced as the final stage, and this has cut costs even further. Mr Hicks comments on the company's approach:

> Water is the most abundant solvent known to man, yet it is regarded as inefficient for the removal of organic contamination . . . to improve the [cleaning capacity] of water . . . extremely aggressive additives were used. We have now developed technology which, using a systematic approach, allows [the use of] milder and more environmentally acceptable additives . . . Although some of the more basic cleaning tasks can be carried out in semi-aqueous systems, we see them as an intermediate step in attaining the goals of fully neutral aqueous products systems. We believe that industrial chemical cleaning processes should forthwith be seen as an integrated system, not a one-shot chemical. This is why we have spent so much time and money creating system designs, applications expertise and improved material handling. It is our ultimate corporate aim to eliminate the need for all volatile solvents and provide products which are engineered to minimise the impact on the environment [and create] a safer, more pleasant workplace.

6.5

Waste, by-products, and new business opportunities

A CASE HISTORY FROM WASTE SERVICE NSW

Waste collected by Sydney's municipal councils is managed and operated by a government agency, Waste Service NSW, which contracts out the operation of its facilities to a variety of waste service providers. Waste Service NSW operates in four waste business areas. Solid waste is managed via a Sydney-wide network of six waste recycling and transfer stations, and five landfill depots and recycling centres for council, commercial, industrial, and household customers. These facilities provide an integrated system for recycling, processing and disposing of waste materials.

Based on the anticipated population increase in the Sydney Region, by the year 2000 3.8 million tonnes of solid waste will be deposited in the Sydney metropolitan area if there is no variation to the per capita waste generation rate. The NSW Government aims to reduce by half the amount of solid waste disposed to landfill by that year. This policy has been strongly supported by the Joint Select Committee Upon Waste Management, which unanimously agreed that 50 per cent reduction is the minimum accepted target. To meet this target a maximum of 1.9 million tonnes per year of solid waste can be disposed of as landfill, and the rest will have to be managed some other way. The quantity of waste disposed of to landfill has already been reduced.

Large quantities of materials are being recovered for reuse or recycling. While complete statistics are not available, the amount is believed to be about 1 million tonnes a year. This level is being achieved mainly by collecting commercial waste for recycling and a growing amount of domestic waste as a result of assistance from the Council Recycling Rebate Scheme. Recycling levels are likely to increase further under the present policies, but other initiatives will be required for reaching the 50 per cent diversion target.

Waste Service NSW sees the role of waste management in the context of a world that consists of two environments—the natural one, and the one people have created. Good waste management properly controls the flow of materials that everyday activities impose on the natural environment. An integrated approach means a commitment to reducing the impact that our consumer goods and services make on our natural ecosystems in a way that our community can realistically afford. This fairly pragmatic approach encompasses waste prevention and minimisation, reuse and recycling, materials and energy recovery, disposal, and methane energy recovery.

The Aqueous Waste Plant (AWP) at Lidcombe treats those unavoidable industrial liquid wastes that have no further use, and diverts residues for recycling or secure disposal. In the Sydney region, industry generates up to 200 000 000 litres of these wastes each year. These wastes, which on average contain about 80 per cent water, are mostly oil/water mixtures, paint washings, spent acids and alkalis, interceptor pit pump-outs and a small amount of spent solvents. A combination of conventional processes make the liquid waste safe for discharge into the sewer system. Special waste materials require particular treatment before and during disposal. In 1993 Waste Service NSW processed 41 084 tonnes of special waste including materials such as asbestos, dusty wastes, and customs and quarantine wastes.

Waste Service NSW was one of the earliest government agencies to combine public ownership with private sector operation. All the sites, except the specialised depots at Lidcombe and Castlereagh, are operated by private sector partners with extensive industry expertise. This joint business strategy gives the community the dual assurances of long-term environmental care and operating efficiency. Operating contracts impose strong requirements for onsite environmental care by their partners. For example, at transfer stations, waste is always sealed into transfer trailers at the end of the day. All landfilled waste is immediately compacted and covered with earth. This prevents odours, discourages vermin and seagulls, and keeps leachate and contaminants in check, although leachate is even used sometimes as a fertiliser for revegetation programs. Water spraying and regular litter patrols reduce dust and windblown waste, while gas flares minimise unpleasant odour problems.

Since operations are affected by the weather, it is necessary to gather data on wind patterns, rainfall and other environmental influences. Meteorological stations at the Lucas Heights, Castlereagh, Belrose and Eastern Creek depots aid the design of more accurate environmental controls and gas extraction procedures in the future.

Waste Service NSW achieves one of the world's highest waste compaction rates—greater than 1.1 tonnes per cubic metre. Heavy compaction not only reduces leachate but makes it possible to use the methane produced by the sites as an alternative source of long-term energy. At present the methane energy is supplied to local industry and electricity supply authorities.

Waste Service NSW constructs engineered landfills to ensure that its operations do not damage local environments. The landfills are regarded as bioreactors that use natural processes to achieve more rapid stabilisation. Design and operating practices are modified to suit the special environmental needs of each site. It is only now that this approach is being considered overseas in favour of the former practice of deferring natural processes until some time in the future.

Waste Service NSW puts considerable energy into developing pragmatic ways to improve waste stream management. At the Rockdale waste recycling and transfer station an analysis of the traffic flow led to a trial in which vegetation and wood wastes were collected separately on weekends. The response to the trial was encouraging and a potential 1500 tonnes per annum will now be diverted from landfill.

It is important to remember there may always be some sort of disposal requirement, since a complete conversion to a fully recycling society may be unrealistic. The quantities left over for disposal, however, will become much smaller and therefore much easier to manage. Waste Service NSW envisages that eventually there will be materials recovery facilities (MRFs). Instead of sorting recyclables at home into an ever-increasing number of bins, all these materials will be collected from one bin and separated at a MRF by mechanised and manual sorting, for return to the market. Many landfills will become what are now called Resource Recovery Industry Cluster Parks. Three to five of these parks could be strategically placed next to landfill sites throughout metropolitan Sydney. Each would process 200 000–400 000 tonnes of waste materials per year, with only the residues going to landfill. Specialist industries such as energy recovery plants, wood product, building material and building demolition material recovery plants, would sort out recyclable materials from the waste. It has been estimated that this would boost the economy by $500 million a year, reducing imports and increasing exports.

A significant part of the business capability of Waste Service NSW is its provision of consultancies to all waste management fields. Its valuable technical expertise encompasses everything from transfer station design and composting systems, to in-house recycling schemes and advice on community consultation programs. Waste Service NSW exports its expertise—for example, it helped the Municipal Authority of Hong Kong to upgrade considerably its waste management operations, which service almost 5 million people.

6.6

The Halon Bank—the first of its kind in the world

A CASE HISTORY FROM DR BRENT DAVEY, TECHNICAL DIRECTOR OF THE DEPARTMENT OF ADMINISTRATIVE SERVICES CENTRE FOR ENVIRONMENTAL MANAGEMENT (DASCEM)

Since the discovery of the hole in the ozone layer over Antarctica, there has been widespread concern about the effects of a global loss of ozone from the atmosphere on human health and the environment. Although ozone is only a trace gas (about 0.4 parts per million by volume) it is essential for absorbing solar ultraviolet radiation, which is hazardous to life on Earth. Scientists consider ozone to be the most chemically active trace gas in the atmosphere. Unlike other gases, which are concentrated in the troposphere, about 90 per cent of the ozone occurs in the stratosphere about 15–35 kilometres from Earth's surface.

Up until the mid-1970s ozone concentrations over Antarctica were reasonably stable, but scientists at Halley Bay then found that the ozone levels were progressively declining each spring. By 1987, 50 per cent of the continent had less than half the levels of ozone noted in the early 1970s, and this depletion is now referred to as a 'hole' in the ozone layer. Ozone depletion is largely caused by traces of halogenated substances. One of these is called Halon. Halon (HALogenated hydrocarbON) is not a household word, but when it is released from fire extinguishers it can be up to 16 times more damaging to the ozone layer than chlorofluorocarbons.

Halon is an artificial substance which until recently was widely used in fire suppression systems in building complexes, in extinguishers for homes, boats and caravans, and in explosion suppression systems. Australia has about 4500 tonnes of Halon, most of it in about 2.5 million 'yellow' fire extinguishers (BCF). These cylinders are unstable and must be checked regularly for leaks.

Halon's ability to destroy ozone was not appreciated until the early 1980s. Since then it has been less popular and stocks from decommissioned systems have been accumulated by governments, industry and the community. It became urgent to set up a facility to collect, store and destroy used Halon under controlled conditions to prevent accidental release. There was also a need for Halon reserves because in some cases (e.g. submarines) no alternative is available yet.

Australia is a signatory to the Montreal Protocol on Substances that Deplete the Ozone Layer, and this obligation to protect the ozone layer involves phasing out yellow extinguishers and Halon in general. In 1992 the Federal Government decided to create a National Halon Management Facility, now known as the Halon Bank, to assist the phase-out process. It is the world's first program that actively reduces the volume of our gas emission. Previous efforts to reduce the effects of ozone-depleting substances have largely relied on reducing or restricting their production. The Halon Bank is the first organisation to collect, store and destroy ozone-depleting substances and to speed up the natural regeneration of the ozone layer. For example, Halon molecules are stable and may take 15–20 years to travel up to the ozone layer where they catalyse its destruction. If nothing was done about the problem apart from terminating production now, present Halon levels would not break down much before the year 2060. However, if even 80 per cent of the world's Halon can be collected and destroyed in the next few years, it could be eliminated from the stratosphere by about 2020—a difference of 40 years.

The primary Halon markets are the Commonwealth and state governments, particularly those with large computer installations, large private sector businesses, and government business enterprises. The Halon Bank aims to collect most of Australia's Halon stock by mid-1995. While the Federal Government subsidises the Halon Bank for collecting Halon from small businesses and domestic users, and for storing and destroying it, there are no strong incentives to encourage others to relinquish stock, although it may soon be illegal in some states to own Halon extinguishers.

Each state has an environmental protection authority that is responsible for its own regulations for the use of Halon, and these regulations vary from state to state. South Australia has moved ahead of the others with its ozone protection policy and its government provides a subsidy to bridge the gap between the Federal subsidy and the cost of depositing Halon in the Halon Bank.

The Halon Bank has undertaken to identify and collect Halon holdings throughout Australia and store them in a controlled environment. It intends to decant and then destroy the substance with the 99.99 per cent efficiency recommended by the Montreal Protocol and Commonwealth EPA requirements. The Halon Bank will also ensure that sufficient stocks of recycled Halon will still be available for approved essential users in Australia until an alternative is found. The Bank is appealing to the environmental conscience of those who are holding on to the Halon and will reward those who respond by promoting them as 'responsible corporate citizens'.

The destruction of Halon will be most effectively achieved by Plascon, the plasma process developed in Australia by a research group from the CSIRO Division of Manufacturing Technology in collaboration with Siddons Ramset Limited (SRL). Plascon is compact and cost-effective, since it is only the size of a small shipping container and can treat about half a tonne of Halon per day, relatively inexpensively. It works by breaking down molecules to their constituent elements at plasma temperatures of over 5000°C. This prevents toxic by-products, which are often formed by processes such as incineration.

Plascon is an electric device that generates plasma—the fourth state of matter—as an ionised, electrically neutral gas made up of molecules, atoms, ions and electrons. If plasma is not to discharge itself quickly (as plasma in the form of lightning does) the supply of free electrons must be maintained by adding energy at a temperature of at least 5000°C. This is best achieved by adding an electric current so that the plasma need not depend on oxygen. In principle any gas can be used, because plasma for Halon destruction works by pyrolysis (degradation by heat) rather than incineration (degradation by oxidation).

The Halon is fed under pressure into the core of the incandescent arc between two copper electrodes. The process works on the same principles as an arc welder, but involves incredibly high temperatures—10 000–15 000°C, considerably hotter than the surface of the sun. This extreme heat makes the Halon molecules dissociate into atoms that recombine as safe, non-toxic compounds, like salt and domestic bleach, which can be used as deodorisers for sewers. It is estimated that a single 150 kW Plascon unit could run 24 hours a day (with a shutdown every 100 hours to replace the electrodes) and dispose of roughly half a tonne of Halon daily.

The Halon Bank has been set up on a commercial basis at the DAS Centre for Environmental Management (DASCEM). The Commonwealth is subsidising the cost of collection from small businesses and the community but the public sector, major industries and government business enterprises are expected to pay the costs of responsibly disposing of their Halon at the Halon Bank. The fire protection industry is actively supporting the Halon Bank by collecting Halon and many local fire brigades are acting as local collecting agents throughout the country. The Halon Bank is just one of DASCEM's major service areas. The Centre also provides services for energy management, the strategic removal and elimination of toxic waste, and assessing and decontaminating contaminated sites.

SECTION 7

Our continuing future

7.1

Energy, economics and the greenhouse effect

A case history from the State Electricity Commission of Victoria

Like all major organisations who make a significant impact on the environment, the State Electricity Commission of Victoria (SECV) has a set of environmental goals. These include being recognised as a community leader by responsibly managing the environmental effects of SECV activities, and striving to enhance the quality of the environment of all SECV worksites.

The SECV is committed to a comprehensive, complex program of activities, including: water resource management in the Latrobe River catchment; the control of emissions to the Latrobe Valley airshed; the Greenhouse Action program; phasing out ozone-depleting substances; the control of hazardous materials; and land rehabilitation.

SECV Greenhouse activities are in accordance with a 'no regrets' policy which means that carbon dioxide emissions are reduced in ways that are cost-effective by current business criteria. Overall, Australia contributes about 1.3 per cent of global CO_2 emissions, so its CO_2 contribution to the greenhouse effect is about 0.8 per cent. Although Australia's emissions of CO_2 are small on the global scale, its per capita levels of CO^2 emissions are amongst the world's highest, mainly because of our high standard of living and dependence on coal for energy generation. So in many ways the SECV's activities may have a greater effect on the environment than most other Victorian businesses.

Abundant low-cost brown coal resources are fundamental to Victoria's power supply, but it is imperative that the burning of the coal is energy-efficient and that the ensuing emissions are as clean as possible. The SECV spends a lot of its energy developing and refining energy efficiency and conservation measures and more efficient electricity generation technologies using brown coal. One of its major activities is using nature itself to solve the emission problem. Tree planting is used to offset CO_2 emissions from its power stations, and 80 hectares of trees were planted in 1992–93, mainly in the Latrobe Valley. Over 1 million trees covering 1500 hectares have been planted in the past four years. These plantings remove 40 kilotonnes of CO_2 from the atmosphere each year as well as improving the appearance of the Latrobe Valley and surrounding areas.

The improved efficiency of the entire community's use of energy is an important area where the growth rate of electricity demand, and hence CO_2 emissions, could be reduced. Current programs include energy conservation such as cogeneration and renewable energy incentives, appliance labelling and industrial/commercial energy audits. Many programs, but by no means all, are cost-effective at current tariffs with attractive payback periods. However, it is not always easy to interest the average consumer in such projects, partly because of the low cost of energy at present, and the desire to minimise capital costs. Nor is demand management easy. Programs for achieving various levels of conservation are evaluated in terms of likely success, and costs.

The greater development of cogeneration is a good example of replacing electricity derived from brown coal by using a more CO_2-efficient fuel such as natural gas, and a more thermally-efficient technology. By using a gas turbine, burning natural gas and generating steam from a waste heat recovery boiler (for low grade heat in industry), electricity can be generated at conversion efficiencies of 50–55 per cent and emissions of 400 kilograms of CO_2 per megawatt hour for a small cogeneration plant of a few megawatts. These values compare with efficiencies of less than 30 per cent and emissions greater than 1000 kilograms per megawatt hour for brown coal stations. It is estimated that several hundred megawatts of cogeneration plant could be developed in Victoria. Cogeneration is being promoted in the current Cogeneration and Renewable Energy Incentives Package.

The use of landfill gas (given off from rubbish tips) is a recent development that uses conventional technology. The gas from the tip, mainly methane, is pumped from the ground, cleaned and burned in gas-fired internal combustion engines. Several projects with capacities of up to 6 megawatts are under consideration in the current Cogeneration and Renewable Energy Incentives Package. The development of these projects will continue to be encouraged because of their financial viability and environmental benefits. The overall effect of burning rather than wasting this fuel, and using the energy this produces to replace its coal equivalent, is to decrease the greenhouse contribution of landfill gas by a factor of about 40.

RENEWABLE ENERGY—INFINITE SOURCES OF FUTURE ENERGY?

Renewable energy sources include hydro, solar, wind, wave, biomass, tidal, geothermal and ocean thermal power. The development of these technologies is at various stages—for example, it is fully developed for hydro, and still experimental for ocean thermal energy. The potential contributions in most cases are limited by natural availability.

Hydro-electricity is already being used in the Victorian system. Victoria's total potential is about 15 000 gigowatt hours per annum, of which nearly 2000 gigowatt hours are considered usable. Those considered unusable generally require extensive civil works (which makes them too costly) but much of this resource occurs in the environmentally sensitive alpine and sub-alpine regions—which would be subject to especially rigorous examination, if development were to be permitted at all.

There has been considerable overseas interest in developing reliable technology for generating electricity from waves. The SECV is evaluating the potential for harnessing wave power for electrical energy productions. Wave power is not yet technically proven for commercial operation, but growing world interest in renewable energy sources may well accelerate its development over the next few years. Renewable energy will make an important contribution to any successful greenhouse mitigation strategy.

Solar energy is still much more expensive than wind energy for large-scale generation. At present, solar thermal technology (absorption of heat and subsequent conversion to electricity) is more economic than solar photovoltaic (direct conversion to electricity) for large developments.

7.2

The race for fuel cell electricity generation

A CASE HISTORY FROM THE STRATEGIC INDUSTRY RESEARCH FOUNDATION

A multi-million dollar race is on, to develop a potent, economical and environmentally friendly source of power known as ceramic (solid oxide) fuel cells. These electrochemical devices convert chemical energy directly to electricity. Ceramic fuels are regarded by many experts as the most efficient system for converting fossil fuels to electricity, and they create less pollution than other technologies.

A consortium called Ceramic Fuel Cells Ltd was formed in 1991 to research and develop fuel cell technology for commercial application in Australia and overseas. This technology has the potential to replace many of our present power generation technologies by the turn of the century. Nine leading Australian companies are involved: BHP, CSIRO, the Electricity Trust of South Australia, the Energy Research and Development Corporation, Generation Victoria, Pacific Power, the Queensland Electricity Commission, the State Electricity Commission of Western Australia, and the Strategic Industry Research Foundation.

Fuel cell technology is not new. It was first discovered in 1839 and has been studied intensively by research teams in various countries for many years. It was not until the 1980s that the attempts to make this form of power generation commercially viable made significant advances. Ceramic fuel cells are third generation fuel cells, following on from phosphoric acid and molten carbonate fuel cells. Ceramic fuel cells have many advantages over their predecessors, including substantially lower manufacturing and maintenance costs, and higher efficiency ratings (60 per cent more efficient without heat recovery and possibly 80 per cent with heat recovery).

The operating temperature of the ceramic fuel cell is higher than its predecessors, with a high-quality waste heat of about 1000°C. Unlike phosphoric acid or molten carbonate systems, ceramic fuel cells also offer broad fuel flexibility and minimal emissions. Ceramic fuel cells directly convert fuel and oxidant (in the form of air), into electricity. More specifically, the oxidation-reduction process occurs across an electrolyte, bound on one side by an anode and a cathode on the other. The electrolyte—in this case a ceramic material—conducts oxygen ions.

The fuel—generally hydrogen or carbon monoxide—is applied to the anode whilst air or oxygen is applied to the cathode to generate a current. Natural gas may also be used as the fuel, and this may be a better option as this process produces no carbon deposits. Other advantages over more conventional energy systems include the potential to produce electricity efficiently from several fuel sources, and shorter installation lead times. Fuel cells, like batteries, are modular. The unit cell is likely to be relatively small—in the centimetre range—and the relationship between cell numbers and power output is approximately linear. The fuel cells differ from present systems in two important respects: significant diseconomies from small installations are unlikely; and the mass production of cells and stacks is possible.

The cells have a large market potential and a variety of applications, from large-scale power generation to remote settlement power and communication systems. A comprehensive study by the Strategic Industry Research Foundation, a Melbourne-based company that develops industry R&D programs, found that by the turn of the century in Australia fuel cells could be generating $150–250 million, while the potential markets in the USA, Japan and Europe are likely to exceed $A1 billion.

7.3

The economically and ecologically sustainable future of Australia

PROFESSOR ADRIENNE CLARKE, CHAIRMAN, CSIRO

Professor Adrienne Clarke has held academic teaching positions at the University of Auckland and the University of Melbourne, where she was appointed as Chair in Botany in 1985, and where she is also Director of the Plant Cell Biology Research Centre

Professor Clarke is a part-time Member of the former Executive of the Common wealth Scientific and Industrial Research Organisation (CSIRO), and has been Chairman of the Board of CSIRO since December 1991. She was made an Officer in the General Division of the Order of Australia (AO) in 1991 for her services to science and industry.

The history of European settlement in Australia is one of a struggle with the elements—with the soil, which was infertile, shallow and prone to salting; with fires that threatened life, home and livelihood; with winds that laid landscapes bare; and with water, through either its scarcity, disastrous oversupply or quality. These forces have shaped today's Australian character, its art and literature, its industries—and its science and technology.

For nearly two centuries Australian science, together with the nation's two most important industries—agriculture and mining—have grappled with problems associated with the great geological age and intrinsically low carrying capacity of our landscapes. Typically these are problems resulting from too little water, an erratic climate, and deeply weathered land surfaces that concealed mineral deposits. The prospect of global climate change under an enhanced greenhouse regime raises uncertainties about the environmental challenges the world may have to face.

To place late 20th century trends in perspective, consider that early this century, when western nations were going through what could be called the 'dirty phase' of industrialisation, the world's total population numbered little more than 1.5 billion—scarcely more than the population of modern China. Some time around the end of this century, the world's population will pass six billion. Access to fresh water, soil, clean air and the oceans grows more inequitable daily; each Australian, for example, consumes more energy and materials than 30 Chinese. The USA, with less than 5 per cent of the world's people, emits almost a quarter of the world's greenhouse gases.

Already some of the so-called 'tiger countries' of Asia are recognising that air and water pollution, and resource management problems, are limiting their economic growth. Some of the old cities of Europe and North America have run out of cheap, low-quality land to accommodate waste disposal and sewage treatment plants. These trends present Australian science and its industry partners with opportunities to capitalise on their skills and experience in environment protection, the efficient exploitation of natural resources, waste management, and energy conservation. Along with new technologies for monitoring and protecting the environment, Australia has developed a diverse base of environmental know-how that is itself a valuable, exportable commodity. It is now up to science and industry to seek out areas of opportunity, target them with our best ideas and best technology, and then market them successfully.

Historically, Australia's performance in these areas has fallen below that of most other western nations. For example, CSIRO possesses a great depth and diversity of research capabilities but, until relatively recently, had little experience in turning that research to a commercial advantage. In the 1980s the organisation underwent substantial change, aimed at focusing its research on servicing its customers and end-users, and sharpening its skills in enhancing commercial outcomes.

A key policy change during this period was the Government's decision to require CSIRO and other government research agencies to seek 30 per cent of its research funds from external sources. To prosper, the CSIRO's research divisions, which had traditionally functioned as semi-autonomous entities, were forced to trim or terminate research projects that showed little prospect of commercial return, and to liaise more closely with their client industries. These industries now have a much stronger influence on CSIRO's research direction. The outcome has, on the whole, been very positive, but, as with any change, there have been problems: it is always easier to obtain funding for tactical, problem-oriented research than for strategic programs—the 'deep science' that is most likely to yield novel technologies upon which new industries and new directions might be built.

There is also a category of research that could be called 'national housekeeping'. It is concerned with monitoring aspects of the environment such as water quality, soils, the atmosphere, and the nation's flora and fauna. This type of science is mainly descriptive, but underpins much of the applied science that the organisation does, and provides objective information for national policy-making and for setting environmental standards. It also represents part of Australian science's contribution to the pool of international environmental knowledge. Australia, the only nation to have an entire continent, and an entire continental flora and fauna to itself, has led the world in developing new computer-based techniques for describing, identifying and classifying living species. It is also a world leader in fields such as atmospheric science, oceanography, and environment monitoring and management.

In the world's driest inhabited continent, it is no accident that CSIRO and Australian private industry have developed world-class expertise in areas such as water purification technologies, wastewater monitoring and management, irrigation and catchment hydrology. The Sirofloc water purification technology developed by the Division of Chemicals and Polymers is a prime example of an innovative technology that has achieved commercial success. The Sirofloc process, which uses magnetic iron oxide to remove organic compounds and particulate matter from water, is suitable for purifying drinking water, and for treating sewage. A full-scale plant is now operating at Sydney's Malabar sewage treatment plant. Three Sirofloc water supply plants for removing colour and turbidity from drinking water—the largest of 75 megalitres per day—have been built in Britain, and a demonstration plant has been built in Taiwan.

Chance plays an important role in science, and it was a chance observation by CSIRO and Melbourne University researchers that led to the development of a low-cost, simple sewage-treatment process. Considerable effort has been invested in promoting the process to the People's Republic of China, which is now particularly interested because of its low capital cost and suitability for remote communities. The new process stems from an observation that bacteria which normally break down carbon-containing materials in the presence of oxygen, can also use the oxygen present in nitrates to fuel respiration under anaerobic conditions. By alternating between aerobic and anaerobic conditions, the process can convert the ammonium in sewage to nitrate, then to inert nitrogen that can be safely vented to the atmosphere. This technology is being offered to the international market by a consortium—the Australasian Water Alliance.

Although chance plays a role in innovation, it does not feature in the development and marketing phases of a new technology. Sheer persistence is an essential attribute—and this can only flow from the conviction that one's technology is either unique or the best in the world, and solidly based on science. Two Australian technologies serve as examples: CSIRO's Airtrak; and the water-purification process developed by the Sydney-based company Memtec. They also illustrate the obstacles that can make new technologies flounder before they establish themselves in the marketplace.

Memtec was founded in Sydney ten years ago on a radical new continuous microfiltration technology that employs very fine, hollow, porous polymer fibres that can be flushed by high-pressure air, blown back through the fibres, when they become clogged with particles. It can be used to purify drinking water, to recycle municipal wastewater, or for the tertiary treatment of sewage. The Memtec process achieves removal rates that are better than conventional biological digestion processes, and removes all bacteria, viruses, algae and protozoa. Compactness, and the process control inherent in the treatment, is the big attraction of the system, especially for European cities that cannot find land for conventional sewage farms.

Memtec encountered many problems, both in the domestic and international markets. The water and wastewater industries are understandably conservative, and do not experiment with new technologies before they have been rigorously evaluated. New demands for accountability mean that public sector officials also adopt very conservative attitudes. Memtec found it needed massive quantities of repetitive data to convince engineers, who were influential in decision-making. The solution was eventually obtained by constructing demonstration plants. Memtec is now prospering—it has become an international company with 1000 employees and annual sales of over $150 million.

Commercially successful research often has a long history of basic and strategic research. In the late 1950s CSIRO wool researchers embarked on an intensive research program to maintain the status of wool as the world's premier fashion fibre against the threat of synthetics. A key strategy was to study the basic properties of the wool fibre—its microstructure, its molecular characteristics, its physics, and its processing qualities. This work continues today.

In the 1980s the Division of Wool Technology in Geelong developed Siroscour, a new technique for wool scouring that substantially reduces effluents that require costly treatment before discharge into waterways. In early 1994 the Division announced that it had developed a new wool-dyeing technique, based on insights into how dyes enter and spread inside wool fibres. The Sirolan LTD (Low-Temperature Dyeing) process is being commercialised by ICI in Australia and New Zealand. The Sirolan LTD process saves energy, allows new, deeper colours to be developed, and achieves a near-complete uptake of pigments, reducing the level of effluents containing unused dyestuffs and chemicals.

Australia also has state-of-the-art skills in recombinant DNA technology—genetic engineering. We are beginning to move into an era of high-efficiency, profitable, environmentally benign farming. CSIRO scientists are testing transgenic cotton varieties that will protect themselves against insect pests by the so-called Bt gene, from the naturally occurring insect-killing bacterium, *Bacillus thuringiensis*. Between 40 and 50 per cent of the world's insecticides are sprayed onto cotton crops. On average, just 5 per cent of those insecticides actually reach their target; the rest kill non-target arthropod species (including the natural predators of cotton's pests) or leach into ground and surface water. The new approach to pest control represented by the Bt gene means that a highly specific insecticide can exclusively target leaf-chewing pests. The toxin is harmless to humans and other vertebrates; indeed, it is harmless to most insects. It is safely and completely sequestered inside the plant and, when used in conjunction with computer-based crop and pest management techniques, could yield an enormous reduction in the environmental and health costs of conventional chemical sprays.

Community concern about the purity of our food supply has prompted a significant shift towards organic farming in western nations. Agricultural economists doubt that organic farming, as currently practised, can feed the world's billions. But transgenic organic farming may be a viable compromise—a more environmentally benign form of agriculture. There are some psychological obstacles to the widespread application of recombinant DNA technology to agriculture, particularly relating to its impact on the natural environment. There is a need to inform the community, to involve people in the decision-making, and to emphasise that the coming new age of agriculture promises profound environmental benefits that must be weighed against a rational assessment of its risks.

Asked to nominate the biggest threat to the global environment, many experts in western nations might nominate the enhanced greenhouse effect. As yet, it remains only a potential threat, although for most climatologists the question is one of 'when' rather than 'if'. This is related to what is in my view the most important problem, overpopulation. The rapid industrialisation of Asia, and the rising energy consumption that inevitably accompanies economic growth and improved living standards, add urgency to the need to know. While we wait for the verdict, prudence demands that we look at every aspect of energy use to see where savings might be made.

Smelting is an ancient and energy-inefficient technology; CSIRO's new Sirosmelt direct-reduction smelting technique, and the comparable Hi-smelt process developed by CRA, both offer substantial gains in energy efficiency and reduce wastage. Not only are they more environmentally benign than conventional smelting, but they are cheaper. Both techniques are based on a detailed understanding of the fluid dynamics of molten baths.

These few examples show that Australia has a substantial reservoir of talent and experience in environment-related research, and that there is a growing international demand for new ideas, new systems and new technologies for protecting and monitoring the environment. New commercial opportunities are opening up, particularly in Asia, but also in the developed nations of North America and Europe. The challenge now is for Australia to match its scientific and technological expertise with the imagination, persistence and business acumen needed to capitalise on them.

7.4

A blueprint for green print

AMANDA ARMSTRONG, 'GREEN AND PRACTICAL',
RADIO NATIONAL SCIENCE UNIT

Amanda Armstrong is the executive producer of Radio National's weekly environment program, 'Green and Practical', which she began with the late Dr Peter Hunt in 1991. 'Green and Practical' has won gold and silver medals at the New York International Radio Festival in 1991 and 1992, and is one of Australia's most comprehensive media programs on the environment. It covers international, national and local issues, discusses practical solutions to environmental problems, and presents community initiatives.

Accuracy, clarity, balance and objectivity apply to environmental reporting just as they do to any other type of reporting. But several issues have affected these objectives. One that is crucial is—who sets the environmental agenda? It is fair to say that the environment groups make most of the running here. They have a vested interest in sending the faxes and making the phonecalls, and if there is sufficient interest the media follows. The time and money constraints of producing programs and publications encourage environmental groups to do the groundwork, especially for the electronic media which relies on a televisual 'event' for the basis of a story.

Sources of information is another area that should be evaluated. The environment movement has become increasingly specialised in the last few years, especially in public relations and media skills. This sophistication is a good thing, but it is important that journalists go beyond the vested interests and the obvious sources such as previous articles and interviews, to talk to people who have first-hand knowledge. This not only makes for more interesting reporting but adds a richer understanding and perspective to the issue.

Then there is the issue of complexity. Many environmental issues are not only complex—from a reporting aspect they can be dry as dust. From a journalistic point of view this means material bound for the 'too hard basket'. Environmental issues encompass science, economics, politics and social values, and this is further complicated by the fact that for most environmental problems there are few concrete answers.

It is easy to become frustrated by scientists' inability to talk in absolute terms, to make bold statements and to predict outcomes. Science is all about theories that change, about doubt and asking questions—not necessarily answering them. To a journalist who wants to know if A will cause B this can be very frustrating since so often the answer is 'It depends'.

Like financial journalism, the language of the environment is becoming increasingly inaccessible and jargonistic. Another challenge for journalists is cutting through the 'greenspeak' which has become all too persuasive. The 'green marketing' of companies ends up barely distinguishable from conservationist material. No matter who you talk to they all say they are 'doing their best for the environment'.

A survey by ARM Quantum [a market research company in Melbourne] on attitudes to the environment found that only about 12 per cent of the people interviewed had any idea what 'sustainability' really meant. This word is now used all the time, and it is supposed to form the basis of future economic development in Australia. Even among the environmental 'in-crowd' there is disagreement on what 'sustainability' actually means. Other contested terms include 'biological diversity', 'intractable waste', 'the precautionary principle', 'old growth forests', to name but a few. Probably one of the most offending phrases that desperately needs replacement is 'environmentally friendly'.

Conflict is what gives the spice to reporting. It is the tension in a particular issue that creates interest. On a simple level this is about putting a 'baddie' up against a 'goodie'. In a way it has been easy and simplistic to put crusading conservationists up against big, bad business. But what is needed is more investigation and questioning: Does environmentalism always cost jobs? Who causes more damage to land—feral animals or humans? Do many of the primary industries really have more genuinely vested interest in our resources long-term than conservation groups do? Environmental journalists need to explore the merits of each particular issue rather than seeking out polarised points of view. This requires research. It is only then that they can draw their own conclusions.

There is no substitute for rigorous, critical journalism in this field, as in any other. The media seems to take the mistaken view that, because at a superficial level the environment is about trees, cuddly and furry animals and quality of life issues, it is a bit soft. Maybe even a bit 'wet' and 'worthy'. This is unfortunate as in the public's eye it is among the top three issues of concern.

Participants

Wildlife Management International Pty Ltd
C/- Grahame Webb
Berrimah Research Farm
Berrimah, NT
PO Box 38151
Winnellie, NT 0821
Phone: (089) 89 2355
Fax: (089) 47 0678

Australian Conservation Foundation
340 Gore Street
Fitzroy, Vic. 3065
Phone: (03) 416 1166
Fax: (03) 416 0767

National Association of Forest Industries Ltd
Forest Industries House
24 Napier Close
Deakin, ACT 2600
Phone: (06) 285 3833
Fax: (06) 285 3855

Kyocera Electronics Australia Pty Ltd
Level 1, 22 Atchison Street
St Leonards, NSW 2065
Phone: (02) 906 8075
Fax: (02) 906 8153

CRA Limited
External Affairs Department
55 Collins Street
Melbourne, Vic. 3001
Phone. (03) 283 3333
Fax: (03) 283 3812

Waste Service NSW
Zenith Centre
Level 4, 821 Pacific Highway
Chatswood, NSW 2057
PO Box 699
Chatswood, NSW 2057
Phone: (02) 412 1388
Fax: (02) 411 8634

Melbourne Water
625 Little Collins Street
Melbourne, Vic. 3000
Phone: (03) 620 0221

Environment Protection Authority (EPA)
Olderfleet Buildings
477 Collins Street
Melbourne, Vic. 3000
Phone: (03) 628 5533
Fax: (03) 628 5699

DASCEM—Department of Administrative Services
Centre for Environmental Management
Rialto North Tower
525 Collins Street
Melbourne, Vic. 3000
Phone: (03) 649 4816
Fax: (03) 649 4895

Strategic Industry Research Foundation
11th Floor, 60 Collins Street
Melbourne, Vic. 3000
Phone: (03) 654 7155
Fax: (03) 654 2894

Greening Australia Limited
GPO Box 9868
Canberra, ACT 2601
Phone: (06) 281 8585
Fax: (06) 281 8590

World Wide Fund for Nature
GPO Box 528
Sydney, NSW 2001
Phone: (02) 247 6300
Fax: (02) 247 8778

Kemcor Australia Pty Ltd
228–238 Normanby Road
South Melbourne, Vic. 3205
Phone: (03) 647 0500
Fax: (03) 647 0518

Midhurst Services Pty Ltd
C/- Ian Hore-Lacy
7 Carrick Street
Mont Albert, Vic. 3127
Phone: (03) 899 9537
Fax: (03) 899 9537

Tasman Institute
457 Elizabeth Street
Melbourne, Vic. 3000
Phone: (03) 326 8033
Fax: (03) 326 8002

State Electricity Commission of Victoria
Monash House
15 William Street
Melbourne, Vic. 3000

The Wilderness Society
130 Davey Street
Hobart, Tas. 7000
Phone: (002) 34 9366
Fax: (002) 23 5112

Simsmetal Limited
GPO Box 4155
Sydney, NSW 2001
Phone: (02) 956 9100
Fax: (02) 954 9680

CSIRO Australia
CSIRO Information Network
PO Box 218
Lindfield, NSW 2070
Phone: (02) 413 7528
Fax: (02) 413 7635

RGC Mineral Sands Limited
PO Box 84
Geraldton, WA 6530
Phone: (099) 64 2700
Fax: (099) 64 1317

Renison Goldfields Consolidated Limited
Goldfields House
1 Alfred Street
Sydney, NSW 2000
Phone: (02) 934 8888
Fax: (02) 934 8555

ACI Glass Packaging
36 Burwood Road
Hawthorn, Vic. 3122
Phone: (03) 810 2311
Fax: (03) 818 3307

ACI Petalite
PO Box 6864
Wetherill Park, NSW 2164
Phone: (02) 725 1300
Fax: (02) 725 1312

Pratt Industries
Manufacturing Head Office
150 Oxford Street
Collingwood, Vic. 3066
Phone: (03) 416 1522
Fax: (03) 417 4768

Head Office
Level 42, South Tower Rialto
525 Collins Street
Melbourne, Vic. 3000
Phone: (03) 614 5166
Fax: (03) 629 1073

Association of Liquidpaperboard Carton Manufacturers Inc.
121 Walker Street,
North Sydney, NSW 2060
Phone: (02) 954 4588
Fax: (02) 954 4546

Ford Motor Company of Australia Limited
1735 Sydney Road
Cambellfield, Vic. 3061
Phone: (03) 359 8211
Fax: (03) 359 8200

Australian Centre for Minesite Rehabilitation Research
2643 Moggill Road
Pinjarra Hills, Qld 4069
PO Box 883
Kenmore, Qld 4069
Phone: (07) 212 4554
Fax: (07) 212 4574

ABC Radio National
'Green and Practical'
Amanda Armstrong
GPO Box 9994
Sydney, NSW 2001
Phone: (02) 333 1500
Fax: (02) 333 1414

Western Mining Corporation
PO Box 860K
Melbourne, Vic. 3001
Phone: (03) 685 6000
Fax: (03) 686 3569

Australian Mining Industry Council
PO Box 363
Dickson, ACT 2602
Phone: (06) 279 3600
Fax: (06) 279 3699

Applied Chemicals International Pty Ltd
97 Spring Street
Melbourne, Vic. 3000
Phone: (03) 650 5633
Fax: (03) 650 5898